Advances in Experimental Medicine and Biology

Volume 1102

More information about this series at http://www.springer.com/series/5584

Wan Mohd Aizat • Hoe-Han Goh
Syarul Nataqain Baharum
Editors

Omics Applications for Systems Biology

Springer

Editors
Wan Mohd Aizat
Institute of Systems Biology
Universiti Kebangsaan Malaysia (UKM)
Bangi, Selangor, Malaysia

Hoe-Han Goh
Institute of Systems Biology
Universiti Kebangsaan Malaysia (UKM)
Bangi, Selangor, Malaysia

Syarul Nataqain Baharum
Institute of Systems Biology
Universiti Kebangsaan Malaysia (UKM)
Bangi, Selangor, Malaysia

ISSN 0065-2598 ISSN 2214-8019 (electronic)
Advances in Experimental Medicine and Biology
ISBN 978-3-319-98757-6 ISBN 978-3-319-98758-3 (eBook)
https://doi.org/10.1007/978-3-319-98758-3

Library of Congress Control Number: 2018958960

This Springer imprint is published by the registered company Springer Nature Switzerland AG
The registered company address is: Gewerbestrasse 11, 6330 Cham, Switzerland

Preface

The advent in omics technologies has revolutionised our perspective in modern biology. Traditionally, DNA, RNA, protein and metabolite have been investigated using basic molecular tools that enable only a few of these elements to be characterised at any particular experiment. However, the coming of age in omics technologies, such as sequencing technologies and mass spectrometry, has ramped the development and research in this scientific endeavour. Ultimately, rather than focusing on a single gene, protein or compound, omics platform allows non-biased and thorough investigation of all these elements.

As such, this book is designed to cater to the need to comprehend omics at the most basic level and how this new concept shapes our research, particularly systems and synthetic biology fields. Most reviews and books on this topic have mainly focused on the technicalities and complexities of these omics platforms (either genomics, transcriptomics, proteomics or bioinformatics), impeding readers from wholly understanding their concepts and applications. This book tackles such a gap and will be most beneficial to novices in this area, university students and even researchers. Basic workflow, practical guidance and examples in each omics are also described, such that scientists can properly and effectively design their experimentation. Furthermore, both systems and synthetic biology areas are also detailed in this book, further enhancing readers' understanding of these integrated topics.

Bangi, Malaysia — Wan Mohd Aizat
Bangi, Malaysia — Hoe-Han Goh
Bangi, Malaysia — Syarul Nataqain Baharum

Contents

Notes on Contributors

Wan Mohd Aizat received his bachelor and PhD degrees from the University of Adelaide, Australia, majoring in Biotechnology and Plant Science, respectively. He then assumed a research fellow position at the Institute of Systems Biology (INBIOSIS), Universiti Kebangsaan Malaysia (UKM), Malaysia. His main research interest is in multi-omics study, in particular employing proteomics to decipher protein profiles of various species from plants such as mangosteen and *Persicaria minor*, animals such as rats and commercial meat products, as well as humans. He has also received formal training for various mass spectrometry instruments, including Bruker Q-TOF and Thermo Fisher Orbitrap, as well as conducted proteomics classes and workshops periodically.

Kamalrul Azlan Azizan holds a PhD in Microbial Metabolomics from the Universiti Kebangsaan Malaysia (UKM) and is currently a research fellow of metabolomics at the Institute of Systems Biology (INBIOSIS), UKM. His research interests include plant and microbial metabolomics, data visualisation and multivariate statistical analysis. He works primarily with gas and liquid chromatography mass spectrometry, aiming at the high-throughput analysis of primary and secondary metabolite fingerprinting. His current research focuses on the development and application of chromatography platforms and chemometric data analysis in the field of allelopathy.

Syarul Nataqain Baharum received her PhD in Molecular Biology and Biotechnology from Universiti Putra Malaysia (UPM). She then received her postdoctoral training in the field of metabolomics at the University of Sheffield, United Kingdom. She has been appointed as the senior research fellow at the Institute of Systems Biology (INBIOSIS), Universiti Kebangsaan Malaysia (UKM), Malaysia. Her research focuses on the new insight of analytical and biological perspectives of the metabolomics in the field of systems biology. It is particularly focused on the understanding of secondary metabolite production in local herbs and plants as well as fluxomics studies of *Lactococcus lactis*. Her work has been awarded prestigious awards, including BioInnovation Awards in 2011 and the

Selangor My Innovation Award in 2014. Currently, she is the Principal Investigator of Metabolomics Research Group and the Head of Centre for Genome Analysis and Technology and the Head of Centre for Plant Biotechnology at INBIOSIS, UKM.

Hoe-Han Goh received his bachelor and PhD degrees from the University of Sheffield, United Kingdom, with specialisation in Plant Molecular Systems Biology. His first and current academic position is at the Institute of Systems Biology (INBIOSIS), Universiti Kebangsaan Malaysia (UKM), as a research fellow. He was appointed as the Head of Plant Biotechnology Centre and is currently Head of Centre for Bioinformatics Research. He has established a Plant Functional Genomics Research Group with research focus on elucidating the functional genomics of tropical plants for biomolecular discovery and crop improvement towards human well-being. His research approaches incorporate multi-omics strategy with bioinformatics integration for systems understanding. His major research specialisation is on transcriptomics, with the latest sequencing technology including Illumina and PacBio.

Maizom Hassan is a senior lecturer and a research fellow at the Institute of Systems Biology (INBIOSIS), Universiti Kebangsaan Malaysia (UKM). She received her doctorate at the University of Tottori, Japan. Her research areas of specialisation are the enzymology, biochemistry, proteomics and integrated pest management (IPM). The outcomes of her research have been presented in many seminars, conferences and symposiums and have been published in various reputed journals, including Q1 journals. Dr. Maizom has received several research grants funded by the research university and collaborates with several industries to expand her research outputs for the benefits of the society. Currently, her research group is interested in developing a novel and potential bio-rational pesticide against several important pests in Malaysia by employing proteomics, enzymology, computational biology and molecular biology approaches to understand molecular mechanisms of enzyme inhibitors for selected enzyme and insect responses.

Ismanizan Ismail received his bachelor's degree in Biochemistry at the Universiti Kebangsaan Malaysia (UKM), Malaysia, in 1992. He then took up a post as a research officer at the Malaysian Palm Oil Board (MPOB) working on an oil palm transformation project. In 1994, he pursued his PhD degree at the Edinburgh University, Scotland, United Kingdom, majoring in Plant Cell Biotechnology. Upon completion in 1997, he returned to UKM and was appointed as a lecturer in the Department of Botany, Faculty of Life Sciences. He was promoted to Associate Professor in 2003 and then to a Full Professor (Plant Cell Biotechnology) in 2012. His research interest focuses on the regulation of plant secondary metabolites through green technology. His research group utilises integrated multidisciplinary of life sciences and engineering with the knowledge, skill and technology of all research disciplines. He is also working on the regulation of gene expression in relation to plant responses towards plant stresses, especially looking at the involvement of miRNA molecules. Administratively, he was the Deputy Director of Center for Research and Instrumentation (CRIM), UKM, and is now taking up the post as the Director of Institute of Systems Biology (INBIOSIS), UKM.

Kok-Keong Loke received his bachelor's degree from the Universiti Kebangsaan Malaysia (UKM) majoring in Bioinformatics. He has then served as a junior research fellow at the Institute of Systems Biology (INBIOSIS), UKM. His main research interest is in computational multi-omics database development and in silico analysis pipelines, particularly on plant genomes, transcriptomes and proteomes with main focus towards INBIOSIS model plant *Persicaria minor*.

Chyan Leong Ng is a research fellow at the Institute of Systems Biology (INBIOSIS), Universiti Kebangsaan Malaysia (UKM). He graduated from the Universiti Putra Malaysia (UPM), where he earned a bachelor's in Biotechnology and a master's in Molecular Biology. He completed his PhD in Structural Biology at the York Structural Biology Laboratory, University of York, United Kingdom. He then worked as a postdoctoral fellow at the Medical Research Council, Laboratory of Molecular Biology, Cambridge, United Kingdom. His research focus is on the structural and functional studies of biologically important macromolecules, with a particular interest on proteins that are involved in secondary metabolite biosynthesis, microbial pathogenesis and antibiotic biosynthesis.

Normah Mohd Noor is a Founding Director of the Institute of Systems Biology, Universiti Kebangsaan Malaysia (UKM). Her groundbreaking work led to the development of a technique for the cryopreservation of embryonic axes of rubber, representing seminal work in the cryopreservation of recalcitrant seeds. Her research on cryopreservation and conservation of tropical recalcitrant fruit species, particularly of Garcinia, Citrus and Nephelium, has led to the development of techniques for the long-term conservation and micropropagation of these economically (and traditionally) important species. Studies in her laboratory elucidated in vitro culture techniques for a wide array of tropical fruit. Her research continues to understand recalcitrant seed behaviour by using a systems biology approach. Prof. Normah has expanded research in her laboratory to include the broad areas of genetic diversity, developmental biology, molecular biology and metabolomics. These are the newest cutting-edge areas for cryopreservation and genetic resources preservation.

Ahmad Bazli Ramzi received his bachelor and PhD degrees from the University of Malaya (UM), Malaysia, and Korea University, South Korea, respectively, majoring in Biotechnology. He assumed postdoctoral fellow position and currently works as a research fellow at the Institute of Systems Biology (INBIOSIS), Universiti Kebangsaan Malaysia (UKM). His present research aims at advancing metabolic engineering, systems and synthetic biology approaches using microbial platforms for bioproducts and biodiagnostics development. He is actively working in promoting biotechnological advances and synthetic biology research in Malaysia through academic courses and research conferences, including as a plenary speaker at the 12th Malaysia International Genetics Congress (MiGC12).

Abbreviations

2DGE	Two-dimensional gel electrophoresis
ABA	Abscisic acid
AEDA	Aroma extraction dilution analysis
AFLP	Amplified fragment length polymorphism
ALA	5-Aminolevulinic acid
ATP	Adenosine triphosphate
ATR	Attenuated total reflectance
BioE	Biological engineering
BLAST	Basic Local Alignment Search Tool
BSA	Bovine serum albumin
CAD	Computer-aided design
cDNA	Complementary DNA
CDS	Coding sequence
CE	Capillary electrophoresis
COG	Clusters of orthologous groups
CRISPR	Clustered regularly interspaced short palindromic repeats
DA	Discriminant analysis
DBG	De Bruijn graphs
DEG	Differentially expressed gene
DIGE	Differential gel electrophoresis
DNA	Deoxyribonucleic acid
DPPH	2,2-Diphenyl-1-picrylhydrazyl
ELISA	Enzyme-linked immunosorbent assay
ENCODE	Encyclopedia Of DNA Elements
ERCC	External RNA Controls Consortium
ESI	Electrospray ionisation
EST	Expressed sequence tag
FBA	Flux balance analysis
FID	Flame ionisation detector
FPKM	Fragments per kilobase million
FT	Fourier transform

GABA	Gamma-aminobutyric acid
GC	Gas chromatography
GC-MS/O	GC-MS/olfactometry
GEM	Genome-scale metabolic model
GO	Gene ontology
GRAS	Generally recognised as safe
GRN	Gene regulatory network
GSEA	Gene set enrichment analysis
GWAS	Genome-wide association studies
HCC	Hepatocellular carcinoma
HPLC	High pressure liquid chromatography
HS	Headspace
ICAT	Isotope-coded affinity tagging
Indels	Insertions and deletions
IPA	Ingenuity Pathway Analysis
IR	Infrared
Iso-Seq	Isoform sequencing
iTRAQ	Isobaric tags for relative and absolute quantitation
KEGG	Kyoto Encyclopedia of Genes and Genomes
LAB	Lactic acid bacteria
LC	Liquid chromatography
LINE	Long interspersed nuclear element
m/z	Mass-to-charge ratio
MALDI	Matrix-assisted laser desorption/ionisation
MCF	Methyl chloroformate
MeJA	Methyl jasmonate
MIAPE	Minimal Information About a Proteomics Experiment
miRNA	MicroRNA
mRNA	Messenger RNA
MRSA	Methicillin-resistant *Staphylococcus aureus*
MS	Mass spectrometry
MSEA	Metabolite set enrichment analysis
MudPIT	Multidimensional protein identification technology
MVA	Multivariate analysis
NCBI	National Center for Biotechnology Information
NGS	Next-generation sequencing
NIST	National Institute of Standards and Technology
NMR	Nuclear magnetic resonance
NR	Non-redundant
OLS	Overlap-layout-consensus
OPLS	Orthogonal partial least squares
P4	Predictive, preventive, personalised, and participatory
PCA	Principal component analysis
pI	Isoelectric point
PIT	Proteomics-informed by transcriptomics

PLS	Partial least squares
PMF	Peptide mass fingerprinting
polyA	Polyadenylation
PTM	Post-translational modification
QC	Quality control
qTOF	Quadrupole time-of-flight
RBS	Ribosome binding site
RIN	RNA integrity number
RNA	Ribonucleic acid
RNAi	RNA interference
RNA-seq	RNA-sequencing
RPKM	Reads per kilobase million
rRNA	Ribosomal RNA
RSBP	Registry of Standard Biological Parts
Rt	Retention time
SBOL	Synthetic Biology Open Language
SDS-PAGE	Sodium dodecyl sulphate polyacrylamide gel electrophoresis
SELDI	Surface-enhanced laser desorption/ionisation
SILAC	Stable isotope labelling by/with amino acids in cell culture
SNP	Single nucleotide polymorphism
SPME	Solid phase microextraction
SPR	Surface plasmon resonance
SWATH-MS	Sequential window acquisition of all theoretical mass spectra
TAG	Triacylglycerol
TALEN	Transcription activator-like effector nuclease
TCA	Trichloroacetic acid
TCA	Tricarboxylic acid
TCM	Traditional Chinese medicine
TMS	Trimethylsilyl
TPM	Transcripts per kilobase million
tRNA	Transfer RNA
UPLC	Ultra-performance liquid chromatography
VOC	Volatile organic compound
WEGO	Web Gene Ontology Annotation Plot

List of Figures

List of Tables

Chapter 1
Recent Development in Omics Studies

Wan Mohd Aizat, Ismanizan Ismail, and Normah Mohd Noor

Abstract The central dogma of molecular biology (DNA, RNA, protein and metabolite) has engraved our understanding of genetics in all living organisms. While the concept has been embraced for many decades, the development of high-throughput technologies particularly omics (genomics, transcriptomics, proteomics and metabolomics) has revolutionised the field to incorporate big data analysis including bioinformatics and systems biology as well as synthetic biology area. These omics approaches as well as systems and synthetic biology areas are now increasingly popular as seen by the growing numbers of publication throughout the years. Several journals which have published most of these related fields are also listed in this chapter to overview their impact and target journals.

Keywords Genomics · Metabolomics · Molecular biology · Proteomics · Systems biology · Transcriptomics

1.1 The Central Dogma of Molecular Biology and Beyond

The central dogma of molecular biology states that the genetic materials of all living being are encoded by their unique DNA sequences (also called gene), transcribed to RNA (also called transcript) and subsequently translated to proteins as the main catalytic entities (Fig. 1.1). Such concept introduced in the early 1950s certainly has been the main limelight for various research in the world, spanning all three domains of life: eukaryotes, bacteria and archea [1]. Watson and Crick were awarded the Nobel Prize in 1962 for such revolutionary insights (not to be forgotten the contribution by Rosalind Franklin who produced the X-ray of DNA), and later, Holley, Khorana and Nirenberg also won the prize in 1968 for completing the genetic coding of protein synthesis [1]. This traditional concept of molecular biology has since then been improved with the addition of metabolite at the end of the workflow

W. M. Aizat (✉) · I. Ismail · N. M. Noor
Institute of Systems Biology, Universiti Kebangsaan Malaysia (UKM), Bangi, Malaysia
e-mail: wma@ukm.edu.my; maniz@ukm.edu.my; normah@ukm.edu.my

W. M. Aizat et al. (eds.), *Omics Applications for Systems Biology*,
Advances in Experimental Medicine and Biology 1102,
https://doi.org/10.1007/978-3-319-98758-3_1

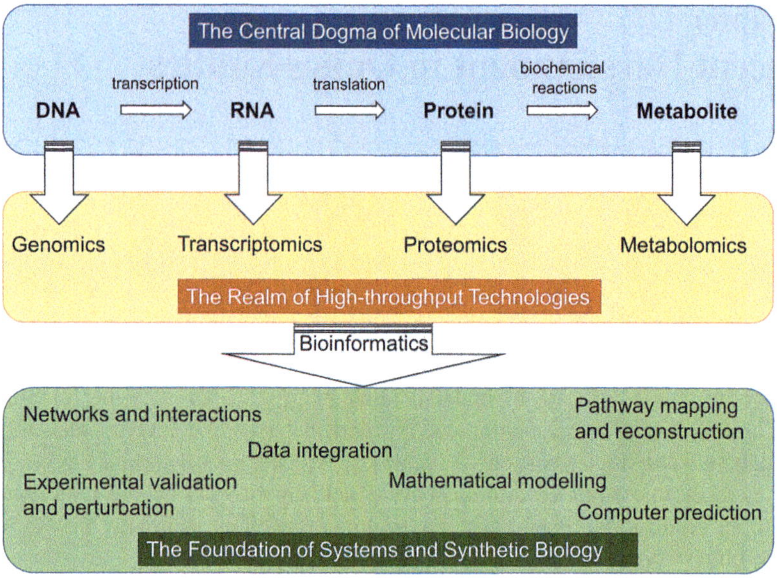

Fig. 1.1 The central dogma of molecular biology comprising of DNA, RNA, protein and metabolite has been greatly investigated by the invention of high-throughput technologies in respective omics (genomics, transcriptomics, proteomics and metabolomics). Such massive data generation requires bioinformatics to analyse and integrate them, ultimately leading to the foundation of systems and synthetic biology

(Fig. 1.1), signifying the importance of the products of biochemical activities catalysed by proteins to influence phenotypic characteristics. One obvious example is how *capsanthin-capsorubin synthase* gene encodes the final enzyme in the biosynthesis of pigmented compounds that gives rise to the red colour of chilies [2, 3]. Without a complete workflow of central dogma, encompassing the downstream metabolites, one would not be able to comprehend the observable characteristics of organisms.

While such a simple relationship between nucleic acids (DNA and RNA) and proteins as well as metabolites has been the mainstream of science research for decades, new formulations of big data science are now taking the stage. Specifically, the introduction of suffixes -omics and -omes has expanded our view in molecular biology. Rather than reductionist approach of focusing on only one or a few genes/proteins, omics approach allows thorough investigation of each facet of molecular biology [4], be it gene (genomics), transcript (transcriptomics), protein (proteomics) and metabolites (metabolomics). While many other omics exist [5], these four omics have been the main themes in molecular biology research, perhaps due to their close relationship with the central dogma itself. Furthermore, these omics approaches are the very foundation of the systems and synthetic biology fields (Fig. 1.1).

1.2 Genomics and Transcriptomics

Investigation of nucleic acids' composition comprising of four bases (adenine, cytosine, guanine and thymine (DNA) or uracil (RNA)) in either genome or transcriptome was greatly facilitated by the development of sequencing platforms. Since the introduction of automated DNA sequencing (1986) as well as the more recent Next - Generation Sequencing (2000) [1, 6], both genomics and transcriptomics areas have exploded with the increasing number of published articles (Figs. 1.2a and 1.3a). Some of the top journals in these areas are *PLOS ONE*, *Nucleic Acids Research* and *BMC Genomics* (Figs. 1.2b and 1.3b). We have combined the great details of the applications and concepts of both omics in Chap. 2 "Functional Genomics".

1.3 Proteomics

Proteins are the workhorses of cellular and biochemical processes. Understanding organism at the level of functional genomics alone may not tell the whole story of how organism functions and responds to environment [7]. As such, proteomics is a vital area to bridge the gap between the more static genome and observable phenotypic characteristics. Evidently, the number of publications related to this area has increased exponentially since the late 1990s (Fig. 1.4a). This is perhaps also due to

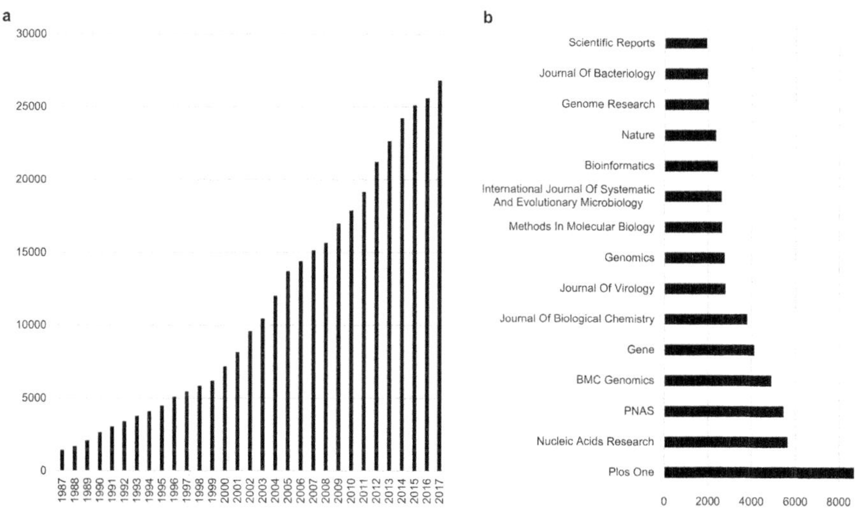

Fig. 1.2 Genomics has become an increasingly attractive research area as shown by the surge of published articles over the years (**a**). The top 15 journals that published the most genomics-related papers of all time are also shown in (**b**). Statistics were obtained from SCOPUS database in July 2018 by searching "Genomic*" in the "Article title, Abstract and Keywords" search field. *PNAS, Proceedings of the National Academy of Sciences of the United States of America*

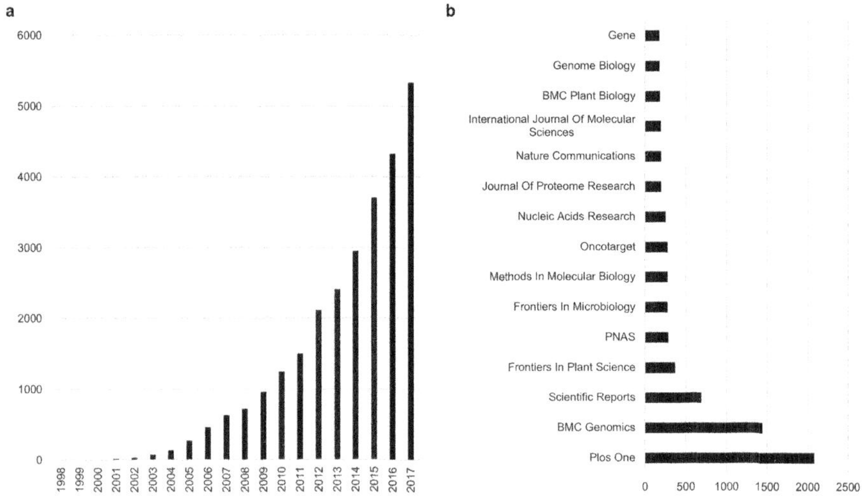

Fig. 1.3 Published articles for transcriptomics studies have steeply increased over the past 20 years (**a**). The top 15 journals that published the most transcriptomics-related papers of all time are also depicted in (**b**). Statistics were obtained from SCOPUS database in July 2018 by searching "Transcriptomic*" in the "Article title, Abstract and Keywords" search field. *PNAS*, *Proceedings of the National Academy of Sciences of the United States of America*

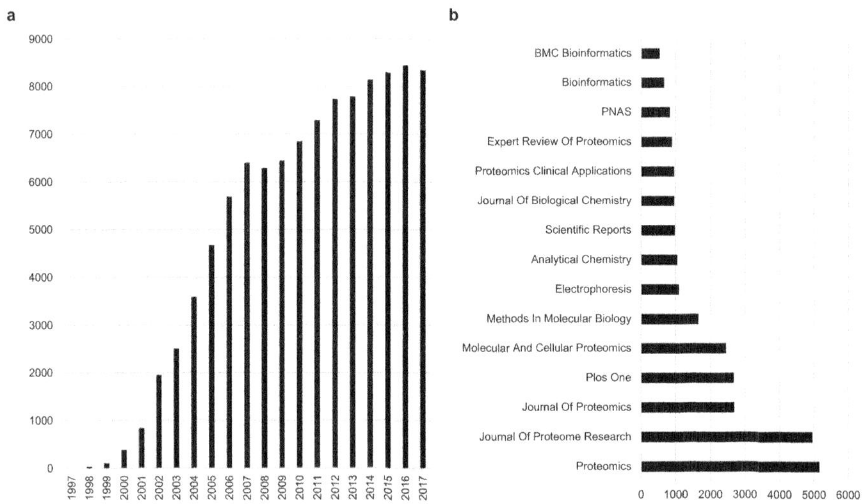

Fig. 1.4 The surge of published articles related to proteomics over the years (**a**) may have been contributed by mass spectrometry development. The top 15 journals that published the most proteomics-related papers of all time are also depicted in (**b**). Statistics were obtained from SCOPUS database in July 2018 by searching "Proteomic*" in the "Article title, Abstract and Keywords" search field. *PNAS*, *Proceedings of the National Academy of Sciences of the United States of America*

the advent and development in mass spectrometry technologies that offer accurate, fast and sensitive methods for protein identification and quantitation [8–10]. The inventors of soft ionisation method in mass spectrometry, John Fenn and Koichi Tanaka, had also been awarded with a Nobel Prize in 2002, suggesting the importance of such system in biomolecular discovery. Some of the key journals that published proteomics papers are *Proteomics*, *Journal of Proteome Research*, *Journal of Proteomics* and *Molecular and Cellular Proteomics* (Fig. 1.4b). We have detailed the applications of proteomics in Chap. 3 "Proteomics in Systems Biology".

1.4 Metabolomics

While proteins may be the workhorses in cells, metabolites also serve as important molecules to be studied in living organism. Metabolomics is a relatively new area compared to the earlier omics, yet its significance cannot be denied. Similar to proteomics, metabolomics also relies on the development of reliable and accurate mass spectrometry systems [11–13]. Such development catalysed the remarkable growth in the paper published in this area since the early 2000s (Fig. 1.5a). Several journals cover this research area extensively such as *Metabolomics*, *PLOS ONE* and *Analytical Chemistry* (Fig. 1.5b). In Chap. 4 "Metabolomics in Systems Biology", this research field is described with emphasis to its applications and current workflow.

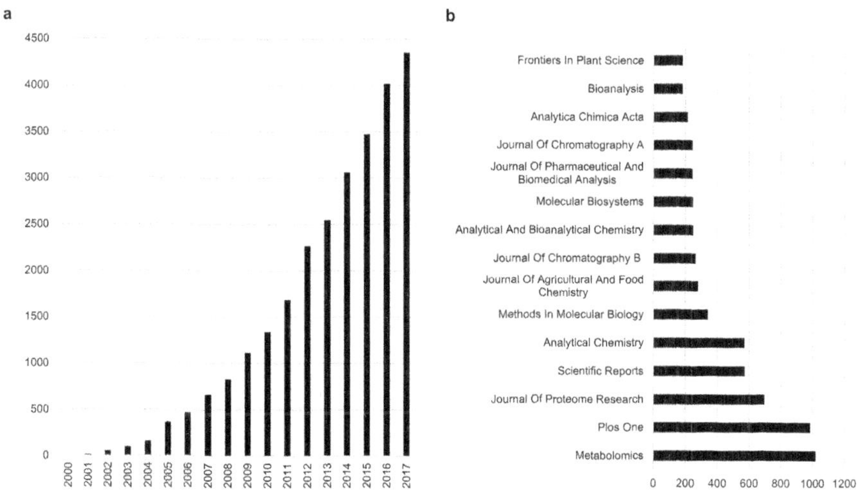

Fig. 1.5 Metabolomics-related articles have been exponentially growing over the years (**a**). The top 15 journals that published the most metabolomics-related papers of all time are shown in (**b**). Statistics were obtained from SCOPUS database in July 2018 by searching "Metabolomic*" in the "Article title, Abstract and Keywords" search field

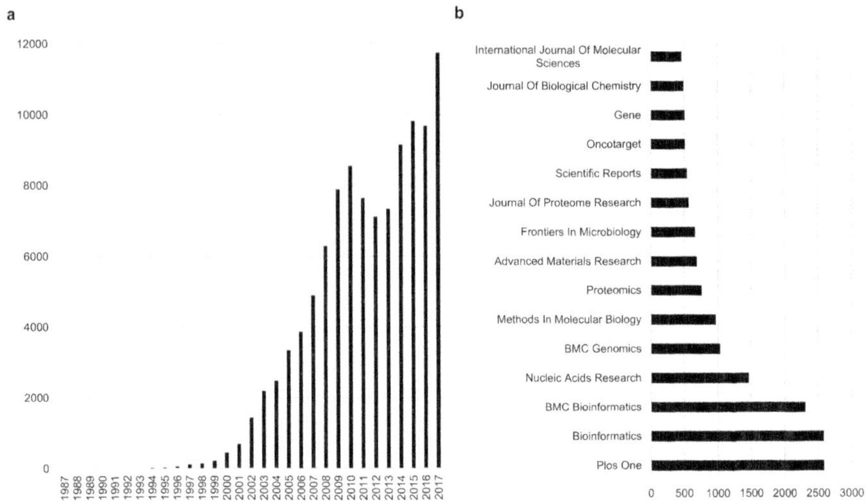

Fig. 1.6 The number of published articles in the area of bioinformatics has surged exponentially over the years (**a**). The top 15 journals that published the most bioinformatics-related papers of all time are indicated in (**b**). Statistics were obtained from SCOPUS database in July 2018 by searching "Bioinformatic*" in the "Article title, Abstract and Keywords" search field

1.5 Bioinformatics

Despite the high-throughput technologies that each omics approach offers, a significant portion of the analysis rely on bioinformatics analysis. Such massively generated data often hamper scientists with "too much" information that may need some sort of computer prediction and analysis to speed up the process [14, 15] (Fig. 1.1). Bioinformatics can be done at each level of omics or in an effort to combine a few omics' results together. This is to "make sense" of the data for scientists to decipher the research question at hand. Furthermore, this research area has certainly benefited with the increasingly powerful computer systems available. Papers related to bioinformatics have reached nearly 12,000 marks in the year 2017 alone, a stark contrast compared to the number of publication in the 1990s (Fig. 1.6a). Some of the journals that actively publish bioinformatics papers include *PLOS ONE*, *Bioinformatics* and *BMC Bioinformatics* (Fig. 1.6b). The concept and applications of bioinformatics in understanding the various omics are detailed in Chap. 5 "Integrative Multi-omics Through Bioinformatics".

1.6 Systems and Synthetic Biology

The different omics concepts have certainly revolutionised the understanding of our modern concept of biology, sparking various technological improvements and breakthrough in science. Systems biology is one area that was initiated by the efforts in understanding cell as a whole rather than its parts [16–18]. This area mainly integrates the different omics to understand the whole process of biochemistry in living being. Systems understanding of biology can be a daunting task, requiring experts in various fields including computer science and wet lab scientists to interrogate and integrate the plethora of information generated (Fig. 1.1). Systems biology ultimately aims to develop mathematical models that explain a biological system by systematically perturbing them in series of experiments [17, 18]. The different approaches in omics will be fed into making the working model to be tested in the experiments. Interestingly, while the papers in systems biology area actively increased from the late 1990s to 2012, a declining trend is seen in recent years (Fig. 1.7a). This may be due to the complexity of the whole topic itself, requiring a more specialised skills to develop the model further [19]. The top journals publishing in this research endeavour are *BMC Systems Biology*, *PLOS ONE*, *Bioinformatics* and *Biosystems* (Fig. 1.7b).

Omics research is also important for the synthetic biology area where biological systems can be designed, interrogated and tested using genetic engineering and genome editing techniques [20]. This research area has actively grown over the last 10 years (Fig. 1.8a), and some of the related journals are *ACS Synthetic Biology*, *Methods in Molecular Biology* and *Proceedings of the National Academy of Sciences of the United States of America* (*PNAS*) (Fig. 1.8b). This topic is detailed in Chap. 6 "Metabolic Engineering and Synthetic Biology".

In this book, each of the chapters is comprised of detailed description of the applications of each omics and its general methodology. We have also expanded the view by incorporating how such omics contribute to the systems biology perspective. While more details on such an infant area of science is duly needed, we reserved the details for now to give broad understanding of this new topic. This is to ensure read-

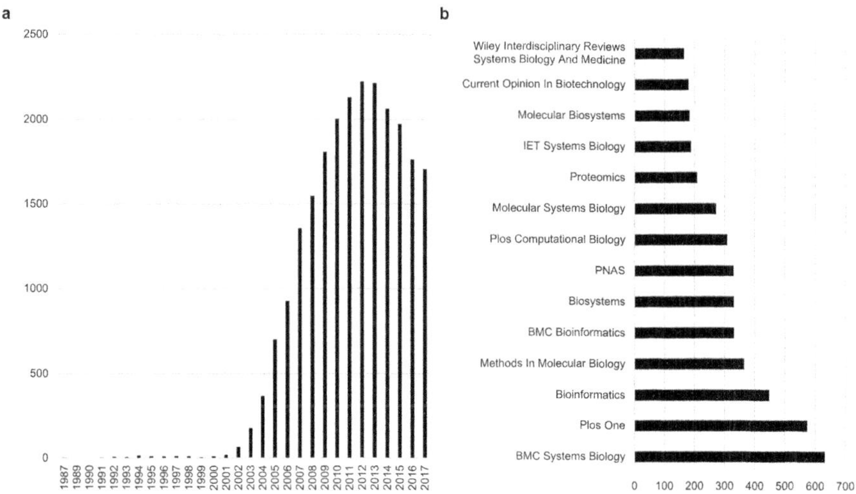

Fig. 1.7 Systems biology has become an increasingly attractive research area as shown by the surge of published articles over the 30-year period (**a**). The top 15 journals that published the most systems biology-related papers of all time are indicated in (**b**). Statistics were obtained from SCOPUS database in July 2018 by searching "Systems biology" in the "Article title, Abstract and Keywords" search field. *PNAS*, *Proceedings of the National Academy of Sciences of the United States of America*

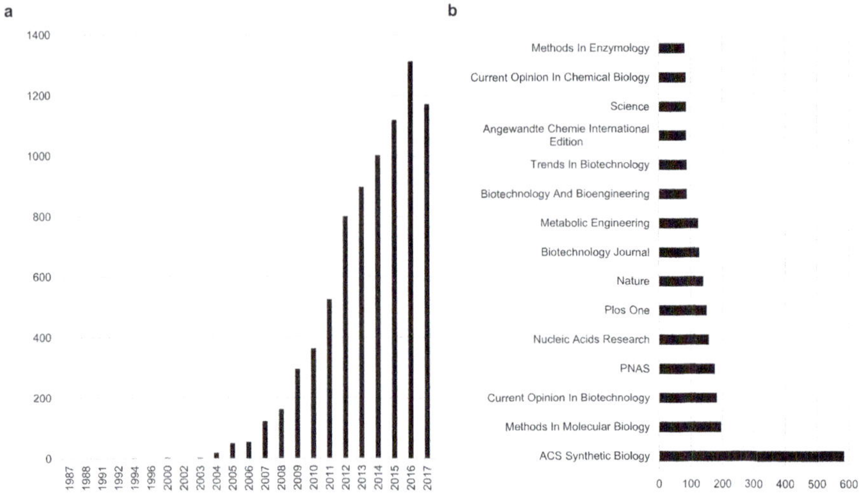

Fig. 1.8 Synthetic biology publications have grown exponentially over the last 30 years (**a**). The top 15 journals that published the most systems biology-related papers of all time are indicated in (**b**). Statistics were obtained from SCOPUS database in July 2018 by searching "Systems biology" in the "Article title, Abstract and Keywords" search field. *PNAS*, *Proceedings of the National Academy of Sciences of the United States of America*

Fig. 1.9 Kesum (*Persicaria minor*) has been utilised in various omics research for systems understanding of this tropical herb

ers grasp the omics concept well and are able to appreciate the level of information generated from these approaches to feed into the systems and synthetic biology research. Furthermore, we have also highlighted our current efforts in understanding tropical herbs such as *Persicaria minor* (syn. *Polygonum minus* or kesum) (Fig. 1.9) and how omics technologies were utilised to characterise this non-model organism.

References

1. Weaver RF (2012) Molecular biology. McGraw Hill, New York
2. Guzman I, Hamby S, Romero J, Bosland PW, O'Connell MA (2010) Variability of carotenoid biosynthesis in orange colored *Capsicum* spp. Plant Sci 179:49–59
3. Ha SH, Kim JB, Park JS, Lee SW, Cho KJ (2007) A comparison of the carotenoid accumulation in *Capsicum* varieties that show different ripening colours: deletion of the capsanthin-capsorubin synthase gene is not a prerequisite for the formation of a yellow pepper. J Exp Bot 58:3135–3144
4. Voit EO (2012) A first course in systems biology. Garland Science, New York
5. Baker M (2013) Big biology: the 'omes puzzle'. http://www.nature.com/news/big-biology-the-omes-puzzle-1.12484
6. Ignacimuthu S (2005) Basic bioinformatics. Alpha Science Int'l Ltd., Oxford
7. Twyman RM (2013) Principles of proteomics. Garland Science, Abingdon
8. Baginsky S (2009) Plant proteomics: concepts, applications, and novel strategies for data interpretation. Mass Spectrom Rev 28:93–120
9. Domon B, Aebersold R (2006) Mass spectrometry and protein analysis. Science 312:212–217
10. Domon B, Aebersold R (2010) Options and considerations when selecting a quantitative proteomics strategy. Nat Biotechnol 28:710–721
11. Dettmer K, Aronov PA, Hammock BD (2007) Mass spectrometry-based metabolomics. Mass Spectrom Rev 26:51–78
12. Patti GJ, Yanes O, Siuzdak G (2012) Innovation: metabolomics: the apogee of the omics trilogy. Nature 13:263
13. Villas-Bôas SG, Mas S, Åkesson M, Smedsgaard J, Nielsen J (2005) Mass spectrometry in metabolome analysis. Mass Spectrom Rev 24:613–646
14. Joyce AR, Palsson BØ (2006) The model organism as a system: integrating 'omics' data sets. Nature 7:198–210
15. Mochida K, Shinozaki K (2011) Advances in omics and bioinformatics tools for systems analyses of plant functions. Plant Cell Physiol 52:2017–2038
16. Gatherer D (2010) So what do we really mean when we say that systems biology is holistic? BMC Syst Biol 4:1–22
17. Ideker T, Galitski T, Hood L (2001) A new approach to decoding life: systems biology. Annu Rev Genomics Hum Genet 2:343–372
18. Kitano H (2002) Systems biology: a brief overview. Science 295:1662–1664
19. Chubukov V, Mukhopadhyay A, Petzold CJ, Keasling JD, Martín HG (2016) Synthetic and systems biology for microbial production of commodity chemicals. Nature 2:16009
20. Cameron DE, Bashor CJ, Collins JJ (2014) A brief history of synthetic biology. Nat Rev Microbiol 12:381–390

Chapter 2
Functional Genomics

Hoe-Han Goh, Chyan Leong Ng, and Kok-Keong Loke

Abstract Functional genomics encompasses diverse disciplines in molecular biology and bioinformatics to comprehend the blueprint, regulation, and expression of genetic elements that define the physiology of an organism. The deluge of sequencing data in the postgenomics era has demanded the involvement of computer scientists and mathematicians to create algorithms, analytical software, and databases for the storage, curation, and analysis of biological big data. In this chapter, we discuss on the concept of functional genomics in the context of systems biology and provide examples of its application in human genetic disease studies, molecular crop improvement, and metagenomics for antibiotic discovery. An overview of transcriptomics workflow and experimental considerations is also introduced. Lastly, we present an in-house case study of transcriptomics analysis of an aromatic herbal plant to understand the effect of elicitation on the biosynthesis of volatile organic compounds.

Keywords Crop genomics · Genomic medicine · Metagenomics · Pharmacogenomics · RNA-Seq · Sequencing · Transcriptomics

2.1 Introduction

Functional genomics is a field of molecular biology that integrates genomic and transcriptomic data to describe gene (and protein) functions and interactions. Genomics is a study of the function and structure of genome, which comprise the complete set of all genes, regulatory sequences, and non-coding regions within an organism's DNA. This discipline in genetics relies on sequencing and bioinformatics approach to sequence, assemble, and analyse all the gene coding and non-coding

H.-H. Goh (✉) · C. L. Ng · K.-K. Loke
Institute of Systems Biology, Universiti Kebangsaan Malaysia (UKM), Bangi, Malaysia
e-mail: gohhh@ukm.edu.my; clng@ukm.edu.my

W. M. Aizat et al. (eds.), *Omics Applications for Systems Biology*,
Advances in Experimental Medicine and Biology 1102,
https://doi.org/10.1007/978-3-319-98758-3_2

sequences and how these genetic components interact to produce an organism and all its functions. Conversely, transcriptomics is the study of the transcriptome—the complete set of RNA transcripts (including mRNA, rRNA, tRNA, and other non-coding RNA) that are produced by the genome, under specific circumstances or in a specific cell—using high-throughput methods. Sometimes, genomics is used as an umbrella term that encompasses genome-wide studies in many subdisciplines, including transcriptomics, proteomics, metabolomics, bioinformatics, systems biology, and synthetic biology. Hence, genomics provides not only a suite of methods and analytical techniques but also a perspective to study an organism as a whole.

On the other hand, functional genomics focus on the dynamic regulation of gene expression and protein-protein interactions, to elucidate DNA function at the levels of genes, transcripts, and proteins in a genome-wide context. The term "genomics" was first coined in 1986 by geneticist Tom Roderick during a meeting on the mapping of human genome, 66 years after the word "genome" was used by the German botanist Hans Winkler [1]. In the year 2000, genome sequences of the first model flowering plant *Arabidopsis thaliana* [2] and insect fruit fly *Drosophila melanogaster* [3] were published. The year after, two independent draft human genome sequences were reported 1 day after another in Feb 2001 [4, 5], followed by the mouse genome in 2002 [6]. The International Human Genome Project initiated in 1990 was officially completed in April 2003, 5 years after sequencing started in 1998, 2 years ahead of schedule. This provides a reference human genome with composite representative sequence derived from several selected individuals of nearly 100 anonymous donors. Since then, more genomes from individuals of different nations were sequenced [7]. The field has advanced to the cataloguing of regulatory elements and epigenomic mapping. These efforts, like genome sequencing, are not done by individual labs but continued as international collaborations, i.e. the ENCODE Consortium [8] and the Roadmap Epigenomics Mapping Consortium [9].

These genome projects not only drove the emergence of new methods for genome-wide investigations but also provided a framework for global views of biology through the advent of sequencing techniques [10]. This propelled myriads of genome sequencing projects of non-model organisms, including the "Genome 10K Project" for vertebrates [11]. Cancer (epi)genomics is another ongoing research hotpot [12]. This is made possible by a parallel development in bioinformatics tools and resources, such as GO tools for the unification of biology [13] and KEGG [14]. High-throughput data generation demands development in large-scale statistical analyses, such as that for genome-wide association studies (GWAS) [15], while clustering has become an integral tool to partition a large dataset into more easily digestible conceptual pieces [16]. Furthermore, visualisation of genome data is of paramount to comprehend emerging patterns [17], such as Integrative Genomics Viewer [18] and Circos [19]. Therefore, genomic transformation of biology into a data-intensive field has recruited many engineers, physicists, mathematicians, and computer scientists into biological research.

2.1.1 *Different Aspects in Genomic Research*

Current field of genomic research has bloomed into diverse scopes from molecular, cellular systems to population level. Molecular genomics like structural genomics [20], glycogenomics [21], toxicogenomics [22], chemogenomics [23], and pharmacogenomics [24] study particular genomic characteristics focused on molecular biology aspects. Cellular genomics like single-cell genomics [25] investigates cellular behaviour in the context of genomics content. Higher level of research scope encompasses complex interactions between multiple genomics such as epigenomics [12], metagenomics [26], comparative genomics [27], phylogenomics [28], GWAS [29], and translational research of genomic medicine [30]. Therefore, genomic research has span across the continuum of basic and applied research, which can be classified into comparative, functional, and translational genomics (Fig. 2.1).

Transcriptomics and proteomics are key parts of functional genomics. These varied genomic platforms allow researchers to address global, general, and specific questions in biology with respect to the genome under study. For example, GWAS have revealed variations in human genome with numerous single nucleotide polymorphisms (SNPs) that are linked to disease risk [31].

2.1.2 *Functional Genomics in the Context of Systems Biology*

The advancement of genomics provided a critical boost to systems biology by facilitating the prediction of complex systems' behaviours, properties, and active processes. For example, the human genomic network permits the gene prediction of the

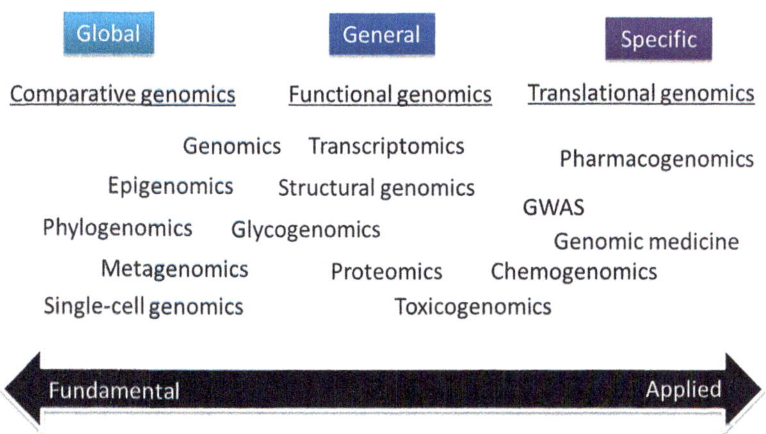

Fig. 2.1 A continuum of diverse research fields in genomics in addressing different levels of biological questions

best drug targets and guides the design of new therapies in treating complex diseases [24]. The ultimate aim is to produce more predictive, preventive, personalised, and participatory (P4) medicine for everyone (further described in Sect. 2.2.1).

Genomics in the broadest sense include both structural and functional aspects. The genome assembly and read mapping are considered structural, whereas the analyses of read abundance and exon usage are functional. These different aspects raise the question "to what extent genomic analyses qualify as systems biology?" [32]. For example, genome assembly is critical to genomic analysis with some of the greatest algorithmic challenges, but the resulting assembly on its own provides little direct insight about the biological system without further analysis.

To address this question, we apply the definition of systems biology as the study of interactions between system parts which involves (i) experimental perturbation, (ii) quantitative measurement, (iii) data integration, and (iv) modelling [33]. For instance, while genome survey solely for genome size and heterozygosity estimation would not fall within the realm of systems biology, the comparative analysis of genome sequences from different cancer cell types to study the genetic variations would qualify. The identification of mutations causing specific cancers represents a systems approach of finding one part in the system which affects the whole system's behaviour.

Reference genome assembly and annotation, as well as comparative genomics, do not shed light on system behaviour on its own, but serve as blueprints for systems-level analysis, such as gene regulatory network (GRN) inference. Functional genomics to study the genome "in action", such as tissue-specific gene expression and the dynamics of transcriptional regulation, are generally within the systems biology framework. Lastly, various genome-wide experiments involving chromatin immunoprecipitation-sequencing (ChIP-seq) interactomics, RNA-seq differentially expressed gene (DEG) analysis, and populational genome variation analysis also fall under the operational definition of systems biology. These analyses typically associate called peaks, expression levels, or variants of specific genes to infer functional enrichment in pathways. Figure 2.2 illustrates how genomics can fit into the context of systems biology through integration with other "omics" platforms.

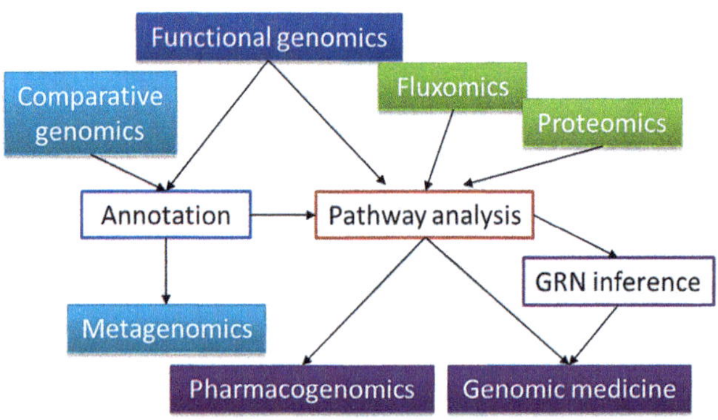

Fig. 2.2 Genomics in the context of systems biology

The integration of genomic data annotation, through functional and comparative genomic analyses, with that of proteomics and fluxomics allows systems-level pathway analysis, which helps in GRN inference. This contributes towards model development in pharmacogenomics and genomic medicine to identify drug targets. Metagenomics expands the study system beyond a single organism towards a community understanding of associated microbes at a functional level.

2.2 Applications of Functional Genomics

Over the past decades, genome sequencing technology has evolved from the first-generation Maxam-Gilbert and Sanger sequencing to current next-generation sequencing (NGS) methods of sequencing by synthesis and single-molecule real-time sequencing [34]. Genome data analyses, such as sequence mapping, assembly, genome annotation, and pathway mapping, have also undergone great advances with the advent of supercomputers, database development, and bioinformatics tools. While genome sequences only provide one-dimensional view on genetic compendium of a cell, when combined with systems biology, it can provide multidimensional insights and understanding on the dynamics of biological processes. In this section, example genomic applications in the studies of human, plants, and microbes are presented.

2.2.1 Comparative Genomics in the Study of Human Genetic Variation

A genome library provides an overview of the genetic makeup of a single organism. Comparison of multiple genome libraries from a single species provides insights into the genetic variation. The 1000 Genomes Project Consortium revealed the genetic variation that encompassed 26 human population and 2,504 individuals throughout five continents [7]. More than 88 million different variants were found, with approximately 96% of them represent SNPs followed by short insertions and deletions (indels) and structural variants. The comparative genomics analysis also reported that every individual harbours four to five million genetic variant sites with more than 99% of them are SNPs.

With the accessibility of genome sequencing, various diseases caused by single-gene mutations (Mendelian or monogenic diseases), such as cystic fibrosis, fragile X syndrome, and Huntington disease, can be easily identified within the genome of individuals or families with risk of inheritance. It is known that many genetic diseases are caused by single nucleotide variants that affect protein function through amino acid substitution [35]. By genome analysis, one can also identify SNPs with indirect association to the disease phenotype and important in the disease development. For instance, many SNPs in the non-coding regions are known to play regulatory roles in gene transcription and expression [36]. Some SNPs found in transcriptional elements have been identified to be associated with β-thalassemia [37], tumour formation [38], melanoma [39], and retinal vasculature defects [40]. Genome analysis has also helped to identify loci which are responsible for disease susceptibility and severity, such as for type 2 diabetes, coronary heart disease, systemic lupus erythematosus, hypertriglyceridemia, or even infectious diseases like trypanosomiasis, malaria, and Lassa fever [41, 42].

This comparative genomics has extended our understanding on human population history at the molecular level and helps in linking disease phenotypes with genetic variants. In the foreseeable future, more comprehensive genomics data and analysis will be available to provide guidance in disease prevention, diagnosis, and treatment according to the personal genetic profile, hence moving us towards P4 medicine era.

2.2.2 Plant Functional Genomics for Crop Improvement

Food security is one of the biggest challenges in this century as the current 7.3 billion world population is projected to reach 9.7 billion by 2050 (UN DESA Report 2015). Along with the impact of climate change and water scarcity, crop productions need urgent improvement with new technologies to overcome upcoming challenges. One of such technologies is to apply functional genomics in place of time-consuming and laborious traditional plant breeding.

Since the fully sequenced genome of model plant *Arabidopsis thaliana* [2], more than 100 plant genome sequences are now available, especially important crop plants such as rice [43, 44], maize [45], soybean [46], potato [47], and bread wheat [48]. Other important commodity plants such as African oil palm [49] and rubber [50] are also sequenced. The availability of crop genome information allowed the identification of genes related to important traits, including yield, disease resistance, and stress tolerance. Crop breeders are now able to accelerate hybrid-breeding programme via marker-assisted selection with genotyping-by-sequencing to produce higher-quality crops [51].

One of the recent examples on how genomic studies can benefit the plantation industry is the oil palm genome study of *Elaeis guineensis* and *Elaeis oleifera* [49] with the identification of *MANTLED* locus responsible for the mantled phenotype through epigenome-wide association studies [52]. The methylation of *Karma* long interspersed nuclear element (LINE) retrotransposon was found to be associated with clones of normal fruit yield compared to mantled clones with hypomethylation [52]. It is therefore useful for screening somaclonal epigenetic alterations during in vitro cloning to cull mantling at plantlet stage to prevent commercial and land use losses.

Apart from the development of molecular markers for the selection of superior traits, precision genome engineering is now possible to improve crops using genome editing tools such as transcription activator-like effector nuclease (TALEN) and CRISPR/Cas9 system [53]. Genetically modified crops with disease and pest resistance as well as higher yields could become more acceptable with the recent genome editing techniques to alter specific gene in a more precise manner without introducing foreign DNA. In the year 2014, hexaploid bread wheat with resistance to powdery mildew was generated by simultaneously introducing three targeted mutations into homoeoalleles of mildew resistance locus o (Mlo) using both TALEN and CRISPR/Cas9 technologies [54]. Genome editing with CRISPR/Cas9 is now possible in major crops such as sorghum, rice, maize, and soybean [55–58]. Despite some current technical challenges like the ineffective delivery method [59], genome editing tools which are relatively cheap and easy to apply will revolutionise crop improvement.

2.2.3 Metagenomics: A New Approach for Antibiotic Discovery

The discovery of antibiotic penicillin by Alexander Fleming in 1928 from *Penicillium rubens* mould saved millions of lives in fighting bacterial infection. Since then various kinds of antibiotics were discovered from bacteria during the prosperous age of antibiotic discovery (1940–1990), primarily from cultured bacteria and sensitivity testing with candidate compounds. A new class of antibiotic is hard to find nowadays in current bacterial collections. For instance, the discovery rate for new antibiotics in actinomycetes is only ~1%, which includes the genus *Streptomyces* that contributed to over 80% of antibiotics discovered so far [60]. Furthermore, the pressing need for new antibiotics is compounded by emerging multiresistant pathogenic bacteria.

Metagenomics has revolutionised microbiology in uncultured microflora [26]. It is estimated that 99% of bacterial species are not yet cultured in the laboratory, [61] and 70% of prokaryotic phyla that exist in seawater, freshwater, and soil are unexplored [62, 63]. While studying these uncultured bacteria remains a big challenge, metagenomics has made it possible to decode genomes for not one but a community of bacteria. This genome information opens up a new source for searching novel bioactive metabolite candidates as antibiotics or drugs. For the past two decades, metagenomics approaches have successfully identified novel antibiotics or new derivatives of known antibiotics (Table 2.1). For example, a new antibiotic teixobactin that could inhibit cell wall synthesis without causing resistance in pathogenic bacteria was discovered [64].

More importantly, many antibiotic resistance-related genes and pathways were also discovered through metagenomics studies [71]. Metagenomics also revealed events of antibiotic resistance gene exchange between environmental bacteria and clinical pathogens [72]. Moreover, single-cell and metagenomics studies coupled with bioinformatics, metabolomics, and chemical analysis by Wilson and colleagues discovered unique bioactive chemical compounds (polyketides and peptides) from two uncultivated phylotypes of marine sponges [73]. These findings encourage further exploitation of uncultivated bacteria for drug discovery via systems biology approach.

Table 2.1 New antibiotics or known antibiotic derivatives identified from metagenomics approaches

Antibiotics	Function	Reference
Indirubin	Antimicrobial activity	MacNeil et al. [65]
Turbomycin A and B	Broad-spectrum antibiotic activity against Gram-negative and Gram-positive bacteria	Gillespie et al. [66]
Palmitoylputrescine	Antibacterial activity against *Bacillus subtilis*	Brady and Clardy [67]
Teixobactin	Inhibit bacterial cell wall synthesis	Ling et al. [64]
Antimicrobial peptides buwchitin	Antibacterial activity against *Enterococcus faecalis*	Oyama et al. [68]
Modified chloramphenicol (*Cm*) derivatives 1-Acetyl-3-propanoylchloramphenicol 1-Acetyl-3-butanoylchloramphenicol 3-Butanoyl-1-propanoylchloramphenicol	Inhibitors for methicillin-resistant *Staphylococcus aureus* (MRSA)	Nasrin et al. [69]
Malacidins	Calcium-dependent antibiotics against Gram-positive bacteria	Hover et al. [70]

2.3 Transcriptomics Workflow

Here we describe main aspects of general workflow for transcriptomics analysis based on RNA-sequencing (RNA-seq) with focus on protein-coding mRNA analysis, which include experimental considerations, sequencing approaches, and data analysis. Comprehensive descriptions are beyond the scope of this chapter; readers are recommended to refer to other recent literature for further understanding on other aspects of transcriptomics [74], such as small RNA-seq, degradome-seq [75], translatome-seq [76], targeted RNA-seq [77], single-cell RNA-seq [78], and epi-transcriptomics [79]. News on the latest development in the field of transcriptomics can be obtained by following the RNA-Seq Blog (https://www.rna-seqblog.com/).

2.3.1 Experimental Considerations

Various aspects need to be considered when designing a transcriptomic experiment, some of which are listed in Table 2.2. The first and the most critical aspect is the experimental design in addressing a specific biological question, which not only determine the strategy of all the downstream transcriptomic analyses but also key to the validity of the experimental results. This includes the purpose of the study on whether it is for expression/differential expression study, study of alternative splicing events (gene isoforms), or discovery of novel transcripts.

Table 2.2 Different aspects of transcriptomic experimental considerations

Aspect	Considerations
Experimental design	Purpose of study Expression and differential expression Isoform discovery and alternative expression Novel transcripts (coding/non-coding) Sample acquisition: purity, quantity, quality, storage Number of replicates (biological and technical)
RNA isolation	Conventional methods or kits Qualitative and quantitative measurements QC (gel electrophoresis, RIN) DNase treatment RNA fragmentation
Library construction	Enrichment: Total RNA, polyA+, polyA-, or ribo-depletion library Size selection (before and/or after cDNA synthesis) Small RNAs (microRNAs)? cDNA fragmentation A narrow fragment size distribution or a broad one Amplification 5′ or 3′ mRNA tags Adapter/index barcoding/multiplexing Stranded or non-stranded library Exome captured or un-captured Library normalisation
Sequencing	Sequencing platform to use: cost, accuracy, read length, time, throughput, applications Depth of sequencing Single-end or paired-end sequencing (Illumina) Spike-ins (ERCC)
Analysis	Raw read preprocessing: de-multiplexing, adapter, sequence quality, trimming Sample QC: correlation analysis/PCA Referenced or de novo pipeline Read alignment: mapping, assembly, or both Level of quantification: transcript/isoform, gene, exon Quantification measure: count, RPKM/FPKM, TPM Differential expression analysis: parametric or non-parametric Alternative splicing analysis: splicing event, isoform expression Functional analysis: annotation (NR, Swiss-Prot, Pfam, GO, COG, etc.), overrepresentation/enrichment analysis (GSEA), pathway analysis (IPA, KEGG, etc.) Visualisation: genome browser, sashimi plot, splice graph etc. Integration: pathway mapping, multi-omics, *etc.*

Different purposes require different strategies of RNA-seq. For example, differential expression analysis of gene with high transcript abundance will not require as high sequencing depth as for the analysis of gene expression with very low abundance and the same applied to the profiling of common and rare transcripts. This also influences the number of biological replicates required for the necessary statistical power to achieve the targeted significance level of differentially expressed genes (DEGs). In general, more biological replicates at lower sequencing depth are better than fewer samples at higher depth with equivalent total number of reads [80]. Pseudo-replication of technical replicates should be avoided to identify true DEGs with biological significance. Furthermore, power analysis is recommended to estimate the minimum number of biological replicates under the budget constraint of RNA-seq experiments [81].

In most cases, obtaining sufficient amount of good quality RNA from limited samples can be a bottleneck for increasing the number of biological replicates. This is related to the second aspect of consideration which is RNA isolation in deciding the most suitable method of extracting high-quality (RIN > 8) RNA for cDNA library preparation. Which library preparation method to choose will depend on the purpose of study on whether the target is only the protein-coding RNA (mRNA), non-coding RNA, small RNA (size selection), or total RNA (no selection). Furthermore, conventional mRNA-seq of whole fragmented transcript can be replaced by tag-based sequencing [82] such as 5′ or 3′ end sequencing if only interested in the expression of annotated genes [83]. This also improves gene quantification with higher sensitivity for rare transcript without the need of gene length normalisation during downstream analysis as tagged-based sequencing avoids the situation of longer transcripts crowding out shorter transcripts at low abundance. Most of the cDNA library kit is currently strand-specific to provide a more accurate estimate of transcript expression and genome annotation.

The choice of sequencing platform will be described in the next section, but briefly, it will depend on the purpose of study with considerations on the cost, accuracy, read length, required throughput (depth), and application. For example, Ion Torrent will be suited for transcript variant analysis of model organisms but not suitable for de novo transcriptome profiling. Lastly, downstream analyses after gathering all the sequencing data will be major considerations depending on the biological questions to be answered. In general, these include preprocessing of raw reads by trimming poor reads, QC, and examination of consistency among the biological replicates through correlation or multivariate analysis such as principal component analysis (PCA). This will require some informed judgement on which is the most suitable software/tools or even version to choose based on intended goal. The rule of thumb is to use the most established and up-to-date version of software/tools relevant to the topics of study as described in the latest literature. It is important to report the version of software/tools and database used throughout the data analysis for reproducibility as different versions might influence the outcome of result.

General workflow of transcript reconstruction will be described in Sect. 2.3.2. For more details on transcriptomic analysis and experimental considerations, we can also refer to a good website, RNA-seqlopedia (https://rnaseq.uoregon.edu/), or a recent review [84]. There are many specialised analysis software/tools, either proprietary such as Ingenuity Pathway Analysis (IPA) or open-source Web Gene Ontology Annotation Plot (WEGO), which are developed for various downstream analyses that will require readers to be aware about the latest development in the field as many intensively used software/tools are regularly updated. However, the most up-to-date version is not necessarily the best option; it is therefore important to understand the detailed functions of a software/tool and changes made by the updates. Older versions are generally still available for download if required. Many of these software/tools are available on GitHub, such as Trinity, with detailed documentation. There is currently a trend towards DockerHub, which contains Trinity and all dependent software used for downstream analyses within the Trinity framework. This will allow easy implementation of analysis pipeline in the server or for cloud computing. The reproducibility of analysis will be improved by the recent efforts in moving transcriptomics analysis towards a customisable automated pipeline in cloud computing [85–87].

2.3.2 *Data Acquisition*

Nowadays, transcriptomic analysis is largely dependent on data generated from RNA-sequencing technology, especially for non-model organisms. However, many studies on model organisms such as human and rice still apply the established Affymetrix microarray approach [88] which will not be covered in this section. Some of the current sequencing platforms used in RNA-seq is summarised in Table 2.3. Roche 454 sequencing platform is not included due to the termination of its development for application in the field of transcriptomics. In general, there are two categories of RNA-seq platforms, which are short-read sequencers such as Illumina and Ion Torrent with generally <400 bp and long-read sequencers such as

Table 2.3 Summary on the different state-of-the-art sequencing platforms for transcriptomics

Platform	Read statistics	Description	Model	Advantages
Illumina	25–300 bp (SE/PE) ~7 h–6 days ~0.6–3000 Gb	Sequencing by synthesis, fluorescence	MiSeq NextSeq HiSeq Novaseq	High-throughput deep sequencing for gene quantification and de novo analysis
Ion Torrent	200–400 bp (SE) 2–4.4 h ~0.03–10 Gb	Sequencing by synthesis, proton release	Proton PGM	Rapid sequencing for multiplex-PCR products, especially for mutation analysis and variant detection
Pacific Biosciences	>40 kb (SE or circular consensus) 3 h ~0.5–10 Gb	Single-molecule sequencing by synthesis, real-time fluorescence	RS II Sequel	Long read for full-length transcript sequencing and isoform detection
Oxford Nanopore	Variable depends on library (1D/2D reads), ~10–950 kb ~6 h to few days ~5 Gb	Single-molecule sequencing by synthesis, real-time electrical current	MinION GridION	Long read with minimal library preparation if high quality; direct RNA-seq is possible

Read throughput, run time, and yield per run are based on minimum and maximum values from different models. *PE* paired-end read, *SE* single-end read

Pacific Biosciences (PacBio) and Oxford Nanopore with >10 kb of reads in real time.

The choice of platform will depend on the purpose of experimental study as described above. It is now possible to generate a full-length high-quality transcriptome reference with PacBio isoform sequencing (Iso-Seq) [89] with unprecedented confidence in the identification of novel transcripts and allele-specific gene expression. However, Iso-Seq is still limited by relatively high cost and lower throughput of sufficient read depth for statistical gene expression analysis. On the other hand, if the study is only concerning with the expression of known genes in annotated genome or organism with high-quality transcriptome, 3′ mRNA-sequencing approach that generates only one fragment per transcript such as QuantSeq [90] can greatly increase the depth of sequencing with more sample multiplexing to improve statistical power at lower cost with simplified analysis. Therefore, it is foreseeable that both sequencing platforms will remain relevant for transcriptomics study, respectively, for transcriptomics profiling and differential expression analysis. To date, Illumina platform is still leading the field of transcriptomics in simultaneously providing the most cost-effective option for both applications.

2.3.3 Data Analysis

For RNA-seq data analysis, the major difference which distinguishes between different analysis pipelines depends on the method chosen for transcript reconstruction (Fig. 2.3). There are two main strategies of transcriptome assembly, namely, align-then-assemble or assemble-then-align. These two strategies can be combined to construct a more comprehensive reference transcriptome.

For reference-guided or ab initio assembly, the reference can be based on an annotated genome or a transcriptome reference. aware aligners or gapped mappers are used when mapping reads to genome to account for mapping across exon junctions. Novel transcripts or gene structures can be discovered from non-annotated transcripts, which can then be functionally annotated. If not interested in novel transcripts, reads can be mapped to a reference transcriptome using unspliced or ungapped mappers for more accurate mapping. Transcript identification and quantification can be achieved simultaneously, such as using Cufflinks.

When no reference is available, a reference-free or de novo assembly can be performed using two different types of assembly algorithms, namely, overlap-layout-consensus (OLC) and De Bruijn graphs (DBG). DBG is based on a much faster *k*-mer indexing approach that works well with short reads compared to a more computing intensive OLC that infers consensus sequences based on a layout of all the reads and overlaps information. DBG approach is currently more popular because most RNA-seq studies are using Illumina short-read sequencing. OLC can be useful for longer sequences generated from Sanger or 454 sequencing. The assembled contigs or transcripts are then used as a reference for read mapping to

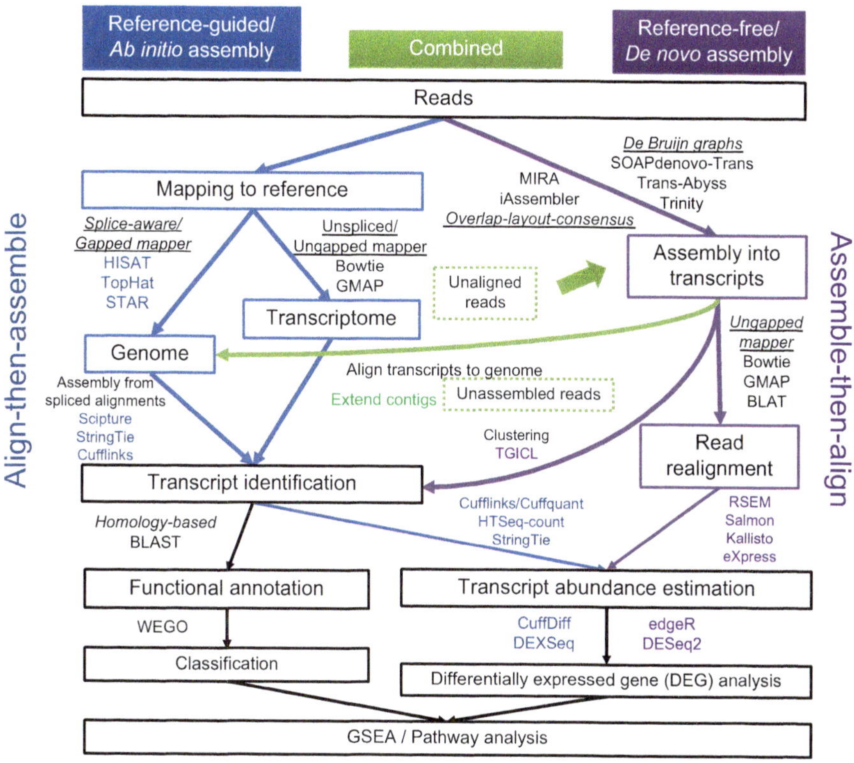

Fig. 2.3 Different analysis pipelines for transcript reconstruction from RNA-seq analysis based on a reference, de novo, or combined approach. Major differences in the approach are underlined. Some of the popular software/tools used for each step of analysis are listed and colour-coded according to different assembly approaches. Common analysis or general tools are in black font

estimate the transcript abundance. This quantification can be estimated at "transcript/isoform" or "gene" level. The transcripts generated from de novo assembly are often subjected to further clustering using software such as TGICL before further analysis.

For a more comprehensive assembly, reads that failed to align to the genome can be de novo assembled, whereas unassembled reads from de novo assembly can be used to scaffold and extend contigs based on the reference genome [91]. This combined approach helps to generate a comprehensive transcriptome that maximises the utilisation of sequencing reads. The final assembled transcriptome will serve as a reference for quantification of expression which can then be subjected to DEG analysis using various statistical software [80] that suit the experimental design or nature of the datasets. For functional annotation, assembled transcripts are BLAST searched against public databases, such as NR and Swiss-Prot, which can then be further categorised according to gene ontology (GO) or clusters of orthologous group (COG). This functional information and results from DEG analysis can then be combined to answer biological questions based on gene set enrichment analysis (GSEA) or pathway analysis for an overview of affected metabolite pathways.

2.4 Case Study: Functional Genomics Study of *Polygonum minus*

Polygonum minus Huds. (syn. *Persicaria minor*) is rich in secondary metabolites with medicinal and pharmaceutical importance [92]. Functional genomics study of *P. minus* started in 2011 with the identification of cDNA for jasmonic acid-responsive genes in root by suppression subtractive hybridisation [93]. The first leaf, stem, and root expressed sequence tag (EST) library was established in 2012 [94]. The is followed by leaf transcriptome profiling of genes induced by salicylic acid and methyl jasmonate (MeJA) through cDNA-amplified fragment length polymorphism (AFLP) approach [95]. All of these studies relied on the low-throughput Sanger sequencing. Recently, de novo RNA-seq using a hybrid NGS approach was taken to construct a more comprehensive transcriptome profile from the leaf and root tissues, respectively, using Illumina sequencing and Roche 454 pyrosequencing [96, 97]. Table 2.4 summarises the statistics from EST library and NGS transcriptome, which shows the great improvement of currently available transcript sequences. Furthermore, DEG analysis of mRNA [98] and small RNA [99] transcriptomes in leaf treated with MeJA can help to understand the effect of elicitation on global gene reprogramming which resulted in the compositional changes of volatile organic compounds (VOCs) [36].

General workflow of RNA-seq analysis, particularly generating transcriptome profile, involves steps in the following order: raw reads preprocessing, filtering and trimming of low-quality reads and contaminant sequences, assembly and clustering, annotation, functional classification, and pathway mapping (Fig. 2.4).

Table 2.4 Statistics of *P. minus* EST library and NGS transcriptome

	EST library [94]	NGS transcriptome [96]
Sequence statistics		
Raw reads	7,292[a]	48,615,711[b]
Average read length (bp)	650	90
Processed reads	5,142	34,365,872
Assembly (*Unigenes*)	4,196	108,541
Functional analysis		
GO terms	2,024	52,796
EC assignment	200	482
KEGG pathway mapped	110	376

[a]Total count of leaf, root, and stem EST clones

[b]Leaf Illumina raw reads were combined with root 454 reads which were clipped to pseudo reads and digital normalised [96]

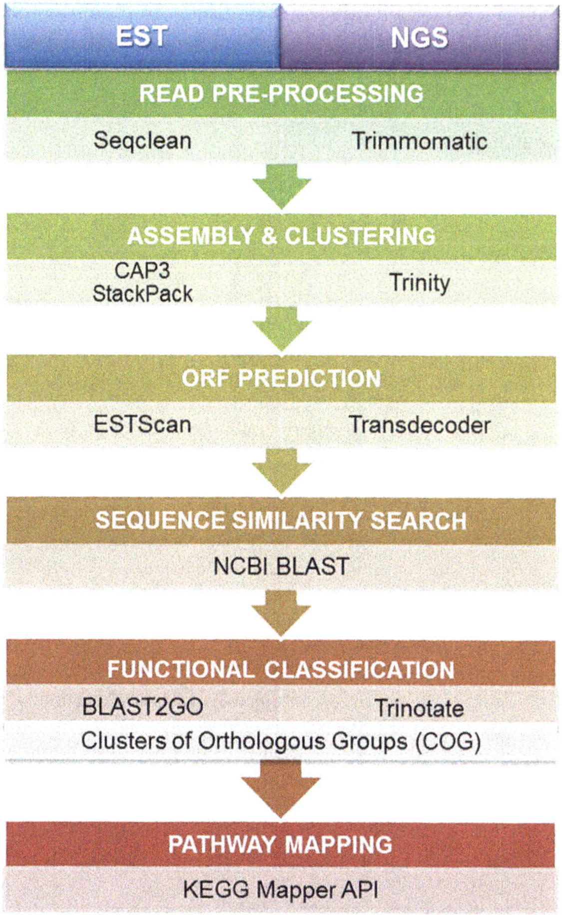

Fig. 2.4 Comparison of analysis workflow in transcriptomic studies of *P. minus*

In EST analysis, a preprocessing step was carried out using Seqclean and then an assembly step using CAP3 and StackPack, followed by open reading frame (ORF) prediction using ESTScan, whereas for RNA-seq analysis, a full Trinity analysis pipeline was followed for de novo assembly. Both required sequence similarity search using NCBI BLAST and functional classification using BLAST2GO based on Gene Ontology and Clusters of Orthologous Groups (COG). Lastly, pathway mapping was performed using KEGG Mapper.

Transcriptome profiling not only contributed to the identification of genes in response to emulated stresses but also allowed the discovery of genes involved in secondary metabolite biosynthesis. Several genes from secondary metabolite biosynthetic pathways were studied. One example is the functional characterisation of sesquiterpene synthase (*PmSTS*) which has been successfully expressed in both *Lactococcus lactis* [100] and *Arabidopsis thaliana* [101]. More recently, a recombinant β-sesquiphellandrene synthase from *P. minus* was expressed and characterised [102]. Furthermore, the transcript sequence database serves as an important reference in proteomics study for protein identification. Increasing availability of genetic information on *P. minus* will help in future exploration of this plant for biotechnological applications.

References

1. Winkler H (1920) Verbreitung und Ursache der Parthenogenesis im Pflanzen- und Tierreiche. Verlag Von Gustav Fischer, Jena
2. Kaul S et al (2000) Analysis of the genome sequence of the flowering plant *Arabidopsis thaliana*. Nature 408:796–815
3. Adams MD et al (2000) The genome sequence of *Drosophila melanogaster*. Science 287:2185–2195
4. Lander ES et al (2001) Initial sequencing and analysis of the human genome. Nature 409:860–921
5. Craig Venter J et al (2001) The sequence of the human genome. Science 291:1304–1351
6. Waterston RH et al (2002) Initial sequencing and comparative analysis of the mouse genome. Nature 420:520–562
7. Auton A et al (2015) A global reference for human genetic variation. Nature 526:68–74
8. Harrow J et al (2012) GENCODE: the reference human genome annotation for the ENCODE project. Genome Res 22:1760–1774
9. Ziller MJ et al (2013) Charting a dynamic DNA methylation landscape of the human genome. Nature 500:477–481
10. Hawkins RD, Hon GC, Ren B (2010) Next-generation genomics: an integrative approach. Nat Rev Genet 11:476–486
11. Koepfli KP, Paten B, O'Brien SJ, Genome KC o S (2015) The genome 10K project: a way forward. Annu Rev Anim Biosci 3:57–111
12. Sandoval J, Esteller M (2012) Cancer epigenomics: beyond genomics. Curr Opin Genet Dev 22:50–55
13. Ashburner M et al (2000) Gene ontology: tool for the unification of biology. Nat Genet 25:25–29
14. Kanehisa M, Goto S, Kawashima S, Okuno Y, Hattori M (2004) The KEGG resource for deciphering the genome. Nucleic Acids Res 32:D277–D280
15. Purcell S et al (2007) PLINK: a tool set for whole-genome association and population-based linkage analyses. Am J Hum Genet 81:559–575
16. Eisen MB, Spellman PT, Brown PO, Botstein D (1998) Cluster analysis and display of genome-wide expression patterns. Proc Natl Acad Sci U S A 95:14863–14868
17. Nielsen CB, Cantor M, Dubchak I, Gordon D, Wang T (2010) Visualizing genomes: techniques and challenges. Nat Methods 7:S5–S15
18. Thorvaldsdóttir H, Robinson JT, Mesirov JP (2013) Integrative Genomics Viewer (IGV): high-performance genomics data visualization and exploration. Brief Bioinform 14:178–192
19. Krzywinski M et al (2009) Circos: an information aesthetic for comparative genomics. Genome Res 19:1639–1645
20. Baker D, Sali A (2001) Protein structure prediction and structural genomics. Science 294:93–96
21. Kersten RD et al (2013) Glycogenomics as a mass spectrometry-guided genome-mining method for microbial glycosylated molecules. Proc Natl Acad Sci U S A 110:E4407–E4416
22. Waters MD, Fostel JM (2004) Toxicogenomics and systems toxicology: aims and prospects. Nat Rev Genet 5:936–948

23. Kanehisa M et al (2006) From genomics to chemical genomics: new developments in KEGG. Nucleic Acids Res 34:D354–D357
24. Hopkins AL (2008) Network pharmacology: the next paradigm in drug discovery. Nat Chem Biol 4:682–690
25. Shapiro E, Biezuner T, Linnarsson S (2013) Single-cell sequencing-based technologies will revolutionize whole-organism science. Nat Rev Genet 14:618–630
26. Handelsman J (2004) Metagenomics: application of genomics to uncultured microorganisms. Microbiol Mol Biol Rev 68:669–685
27. Ellegren H (2008) Comparative genomics and the study of evolution by natural selection. Mol Ecol 17:4586–4596
28. Delsuc F, Brinkmann H, Philippe H (2005) Phylogenomics and the reconstruction of the tree of life. Nat Rev Genet 6:361–375
29. Hirschhorn JN, Daly MJ (2005) Genome-wide association studies for common diseases and complex traits. Nat Rev Genet 6:95–108
30. McCarthy JJ, McLeod HL, Ginsburg GS (2013) Genomic medicine: a decade of successes, challenges, and opportunities. Sci Transl Med 5:189sr4
31. McCarthy MI et al (2008) Genome-wide association studies for complex traits: consensus, uncertainty and challenges. Nat Rev Genet 9:356–369
32. Conesa A, Mortazavi A (2014) The common ground of genomics and systems biology. BMC Syst Biol 8:S1
33. Ideker T, Galitski T, Hood L (2001) A new approach to decoding life: systems biology. Annu Rev Genomics Hum Genet 2:343–372
34. Rhoads A, Au KF (2015) PacBio sequencing and its applications. Genomics Proteomics Bioinformatics 13:278–289
35. Wilson BJ, Nicholls SG (2015) The human genome project, and recent advances in personalized genomics. Risk Manage Healthc Policy 8:9–20
36. Shastry BS (2009) Single nucleotide polymorphisms. Springer, Berlin, pp 3–22
37. Orkin S, Antonarakis S, Kazazian H (1984) Base substitution at position-88 in a beta-thalassemic globin gene. Further evidence for the role of distal promoter element ACACCC. J Biol Chem 259:8679–8681
38. Bond GL et al (2004) A single nucleotide polymorphism in the MDM2 promoter attenuates the p53 tumor suppressor pathway and accelerates tumor formation in humans. Cell 119:591–602
39. Horn S et al (2013) TERT promoter mutations in familial and sporadic melanoma. Science 339:959–961
40. Madelaine R et al (2018) A screen for deeply conserved non-coding GWAS SNPs uncovers a MIR-9-2 functional mutation associated to retinal vasculature defects in human. Nucleic Acids Res 46:3517–3531
41. Janssens ACJW, van Duijn CM (2008) Genome-based prediction of common diseases: advances and prospects. Hum Mol Genet 17:R166–R173
42. Gurdasani D et al (2015) The African genome variation project shapes medical genetics in Africa. Nature 517:327–332
43. Goff SA et al (2002) A draft sequence of the rice genome (*Oryza sativa* L. ssp. *japonica*). Science 296:92–100
44. Yu J et al (2002) A draft sequence of the rice genome (*Oryza sativa* L. ssp. *indica*). Science 296:79–92
45. Schnable PS et al (2009) The B73 maize genome: complexity, diversity, and dynamics. Science 326:1112–1115
46. Schmutz J et al (2010) Genome sequence of the palaeopolyploid soybean. Nature 463:178–183
47. Xu X et al (2011) Genome sequence and analysis of the tuber crop potato. Nature 475:189–195
48. Brenchley R et al (2012) Analysis of the bread wheat genome using whole-genome shotgun sequencing. Nature 491:705–710

49. Singh R et al (2013) Oil palm genome sequence reveals divergence of interfertile species in Old and New worlds. Nature 500:335–339
50. Rahman AYA et al (2013) Draft genome sequence of the rubber tree *Hevea brasiliensis*. BMC Genomics 14:75
51. He J et al (2014) Genotyping-by-sequencing (GBS), an ultimate marker-assisted selection (MAS) tool to accelerate plant breeding. Front Plant Sci 5:484
52. Ong-Abdullah M et al (2015) Loss of Karma transposon methylation underlies the mantled somaclonal variant of oil palm. Nature 525:533
53. Rinaldo AR, Ayliffe M (2015) Gene targeting and editing in crop plants: a new era of precision opportunities. Mol Breed 35:1–15
54. Wang Y et al (2014) Simultaneous editing of three homoeoalleles in hexaploid bread wheat confers heritable resistance to powdery mildew. Nat Biotechnol 32:947–951
55. Jiang W et al (2013) Demonstration of CRISPR/Cas9/sgRNA-mediated targeted gene modification in Arabidopsis, tobacco, sorghum and rice. Nucleic Acids Res 41:e188
56. Lawrenson T et al (2015) Induction of targeted, heritable mutations in barley and *Brassica oleracea* using RNA-guided Cas9 nuclease. Genome Biol 16:258
57. Svitashev S et al (2015) Targeted mutagenesis, precise gene editing and site-specific gene insertion in maize using Cas9 and guide RNA. Plant Physiol:00793.02015, 169(2):931–945
58. Li Z et al (2015) Cas9-guide RNA directed genome editing in soybean. Plant Physiol:00783.02015, 169(2):960–970
59. Gao C (2018) The future of CRISPR technologies in agriculture. Nat Rev Mol Cell Biol 39:1–2
60. Handelsman J, Rondon MR, Brady SF, Clardy J, Goodman RM (1998) Molecular biological access to the chemistry of unknown soil microbes: a new frontier for natural products. Chem Biol 5:R245–R249
61. Davies J (1999) Millennium bugs. Trends Genet 15:M2–M5
62. Rappé MS, Giovannoni SJ (2003) The uncultured microbial majority. Annu Rev Microbiol 57:369–394
63. Achtman M, Wagner M (2008) Microbial diversity and the genetic nature of microbial species. Nat Rev Microbiol 6:431–440
64. Ling LL et al (2015) A new antibiotic kills pathogens without detectable resistance. Nature 517:455
65. MacNeil I et al (2001) Expression and isolation of antimicrobial small molecules from soil DNA libraries. J Mol Microbiol Biotechnol 3:301–308
66. Gillespie DE et al (2002) Isolation of antibiotics turbomycin A and B from a metagenomic library of soil microbial DNA. Appl Environ Microbiol 68:4301–4306
67. Brady SF, Clardy J (2004) Palmitoylputrescine, an antibiotic isolated from the heterologous expression of DNA extracted from bromeliad tank water. J Nat Prod 67:1283–1286
68. Oyama LB et al (2017) Buwchitin: a ruminal peptide with antimicrobial potential against *Enterococcus faecalis*. Front Chem 5:51
69. Nasrin S et al (2018) Chloramphenicol derivatives with antibacterial activity identified by functional metagenomics. J Nat Prod 81:1321
70. Hover BM et al (2018) Culture-independent discovery of the malacidins as calcium-dependent antibiotics with activity against multidrug-resistant Gram-positive pathogens. Nat Microbiol 3:415
71. Li B et al (2015) Metagenomic and network analysis reveal wide distribution and co-occurrence of environmental antibiotic resistance genes. ISME J 9:2490–2502
72. Forsberg KJ et al (2012) The shared antibiotic resistome of soil bacteria and human pathogens. Science 337:1107–1111
73. Wilson MC et al (2014) An environmental bacterial taxon with a large and distinct metabolic repertoire. Nature 506:58–62
74. Hrdlickova R, Toloue M, Tian B (2017) RNA-seq methods for transcriptome analysis. Wiley Interdiscip Rev RNA 8. https://doi.org/10.1002/wrna.1364

75. Ma X, Tang Z, Qin J, Meng Y (2015) The use of high-throughput sequencing methods for plant microRNA research. RNA Biol 12:709–719
76. Aviner R, Geiger T, Elroy-Stein O (2013) PUNCH-P for global translatome profiling: methodology, insights and comparison to other techniques. Translation 1:e27516
77. Li W et al (2015) Comprehensive evaluation of AmpliSeq transcriptome, a novel targeted whole transcriptome RNA sequencing methodology for global gene expression analysis. BMC Genomics 16:1069
78. Saliba A-E, Westermann AJ, Gorski SA, Vogel J (2014) Single-cell RNA-seq: advances and future challenges. Nucleic Acids Res 42:8845–8860
79. Dominissini D (2014) Roadmap to the epitranscriptome. Science 346:1192
80. Lamarre S et al (2018) Optimization of an RNA-seq differential gene expression analysis depending on biological replicate number and library size. Front. Plant Sci. 9:108
81. Ching T, Huang S, Garmire LX (2014) Power analysis and sample size estimation for RNA-seq differential expression. RNA 20:1684–1696
82. de Klerk E, den Dunnen JT, 't Hoen PAC (2014) RNA sequencing: from tag-based profiling to resolving complete transcript structure. Cell Mol Life Sci 71:3537–3551
83. Jamaluddin ND, Mohd Noor N, Goh H-H (2017) Genome-wide transcriptome profiling of *Carica papaya* L. embryogenic callus. Physiol Mol Biol Plants 23:357–368
84. Conesa A et al (2016) A survey of best practices for RNA-seq data analysis. Genome Biol 17:13
85. Griffith M, Walker JR, Spies NC, Ainscough BJ, Griffith OL (2015) Informatics for RNA sequencing: a web resource for analysis on the cloud. PLOS Comput Biol 11:e1004393
86. Nagasaki H et al (2013) DDBJ read annotation pipeline: a cloud computing-based pipeline for high-throughput analysis of next-generation sequencing data. DNA Res 20:383–390
87. Afgan E et al (2018) The galaxy platform for accessible, reproducible and collaborative biomedical analyses: 2018 update. Nucleic Acids Res 46:W537–W544
88. Bair E (2013) Identification of significant features in DNA microarray data. Wiley Interdiscip Rev Comput Stat 5. https://doi.org/10.1002/wics.1260
89. An D, Cao HX, Li C, Humbeck K, Wang W (2018) Isoform sequencing and state-of-art applications for unravelling complexity of plant transcriptomes. Genes 9:43
90. Moll P, Ante M, Seitz A, Reda T (2014) QuantSeq 3′ mRNA sequencing for RNA quantification. Nat Methods 11:972
91. Martin JA, Wang Z (2011) Next-generation transcriptome assembly. Nat Rev Genet 12:671
92. Christapher P, Parasuraman S, Christina J, Asmawi MZ, Vikneswaran M (2015) Review on *Polygonum minus*. Huds, a commonly used food additive in Southeast Asia. Pharm Res 7:1–6
93. Gor MC et al (2011) Identification of cDNAs for jasmonic acid-responsive genes in *Polygonum minus* roots by suppression subtractive hybridization. Acta Physiol Plant 33:283–294
94. Roslan ND et al (2012) Flavonoid biosynthesis genes putatively identified in the aromatic plant *Polygonum minus* via expressed sequences tag (EST) analysis. Int J Mol Sci 13:2692–2706
95. Ee SF et al (2013) Transcriptome profiling of genes induced by salicylic acid and methyl jasmonate in *Polygonum minus*. Mol Biol Rep 40:2231–2241
96. Loke K-K et al (2016) RNA-seq analysis for secondary metabolite pathway gene discovery in *Polygonum minus*. Genomics Data 7:12–13
97. Loke KK et al (2017) Transcriptome analysis of *Polygonum minus* reveals candidate genes involved in important secondary metabolic pathways of phenylpropanoids and flavonoids. Peer J 2017. PeerJ 5:e2938
98. Rahnamaie-Tajadod R, Loke KK, Goh HH, Noor NM (2017) Differential gene expression analysis in *Polygonum minus* leaf upon 24h of methyl jasmonate elicitation. Front Plant Sci 8:109
99. Nazaruddin N et al (2017) Small RNA-seq analysis in response to methyl jasmonate and abscisic acid treatment in *Persicaria minor*. Genomics Data 12:157–158

100. Song AAL et al (2012) Overexpressing 3-hydroxy-3-methylglutaryl coenzyme A reductase (HMGR) in the lactococcal mevalonate pathway for heterologous plant sesquiterpene production. PLOS ONE 7:e52444
101. Ee SF et al (2014) Functional characterization of sesquiterpene synthase from *Polygonum minus*. Sci World J 2014:840592
102. Ker DS et al (2017) Purification and biochemical characterization of recombinant *Persicaria minor* β-sesquiphellandrene synthase. PeerJ 5:e2961

Chapter 3
Proteomics in Systems Biology

Wan Mohd Aizat and Maizom Hassan

Abstract Proteomics is the study of proteins, the workhorses of cells. Proteins can be subjected to various post-translational modifications, making them dynamic to external perturbation. Proteomics can be divided into four areas: sequence, structural, functional and interaction and expression proteomics. These different areas used different instrumentations and have different focuses. For example, sequence and structural proteomics mainly focus on elucidating a particular protein sequence and structure, respectively. Meanwhile, functional and interaction proteomics concentrate on protein function and interaction partners, whereas expression proteomics allows the cataloguing of total proteins in any given samples, hence providing a holistic overview of various proteins in a cell. The application of expression proteomics in cancer and crop research is detailed in this chapter. The general workflow of expression proteomics consisting the use of mass spectrometry instrumentation has also been described, and some examples of proteomics studies are also presented.

Keywords Expression proteomics · Enzyme · Mass spectrometry · Peptide · Protein · Shotgun proteomics

3.1 Introduction

The term "proteomics" was first coined 20 years ago in an effort to define the total proteins encoded by a given genome [1]. Such powerful term remains influential and has since expanded into various fields of research and organisms. The significance of proteomics comes from how important protein is in living cells. The genome of one organism is always static (unless mutation occurs), and yet proteins are expressed based on tissue types, and their expression may be changed upon stimulation of environmental/external conditions. Several post-translational modifications (PTMs)

W. M. Aizat (✉) · M. Hassan
Institute of Systems Biology, Universiti Kebangsaan Malaysia (UKM), Bangi, Malaysia
e-mail: wma@ukm.edu.my; maizom@ukm.edu.my

W. M. Aizat et al. (eds.), *Omics Applications for Systems Biology*,
Advances in Experimental Medicine and Biology 1102,
https://doi.org/10.1007/978-3-319-98758-3_3

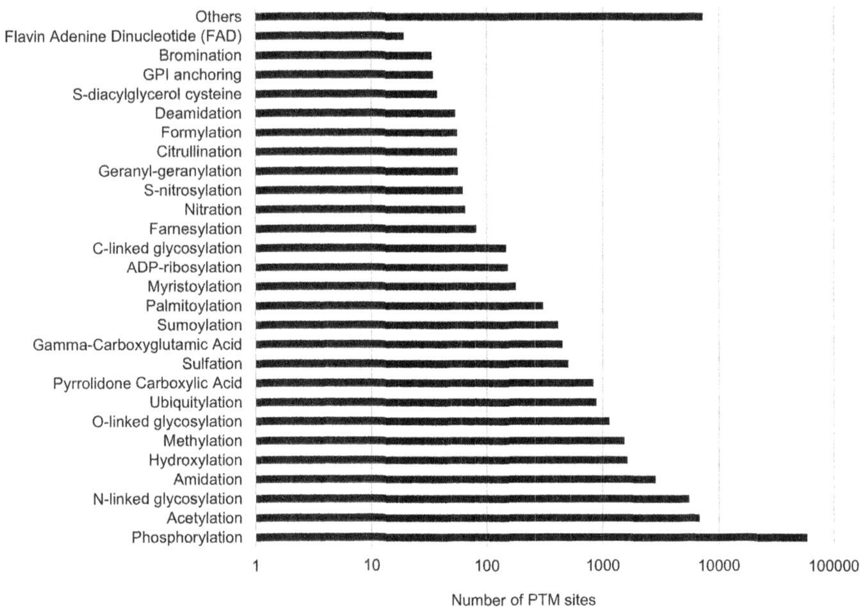

Fig. 3.1 Summary of the number of experimentally observed post-translational modification (PTM) sites documented in the Swiss-Prot database (data obtained from http://selene.princeton.edu/PTMCuration; accessed on July 2018) [3]

(Fig. 3.1) such as phosphorylation, acetylation, glycosylation, amidation, hydroxylation and methylation have also shaped and influenced certain proteins, and hence the level of gene expression may not always correlate with the protein and its activity level [2]. There are estimated more than 200 known PTMs which undeniably increase the proteome complexity of any living being [2, 3]. Furthermore, proteins are known to be the workhorses of cells as they are responsible for various cellular functions such as enzymatic reactions, signaling, gene transcription and translation processes, as well as structural components. This signifies the central role of proteins, and hence the study of proteomics is highly sought for a holistic understanding of cellular regulation.

3.1.1 *Different Aspects of Proteomics*

Generally, proteomics can be categorised into four distinct study areas, namely, "sequence proteomics", "structural proteomics", "functional and interaction proteomics" and "expression proteomics" (Fig. 3.2) [4]. These different proteomics areas tackle different aspects of protein properties, including primary and three-dimensional structure as well as function and protein abundance, respectively.

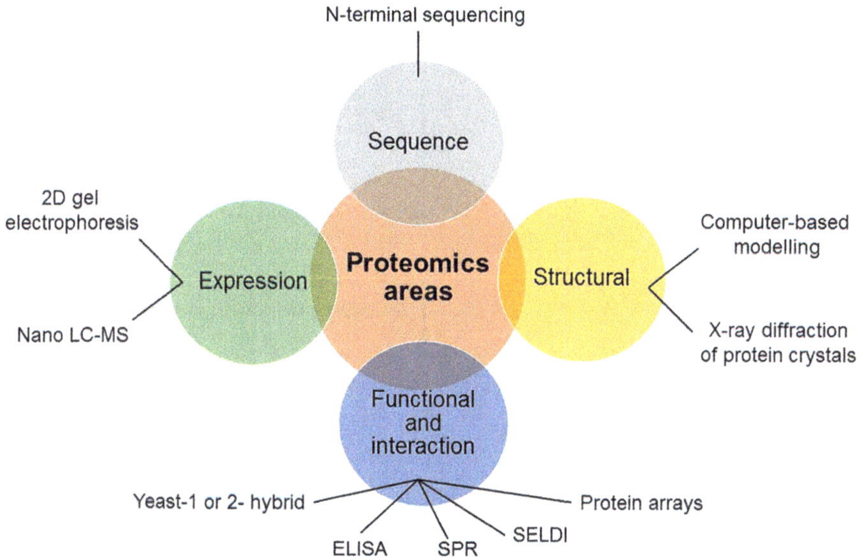

Fig. 3.2 Proteomics can be distinguished into four main aspects which are sequence, structural, functional and interaction as well as expression studies. Each of this aspect is given examples as detailed in text. Surface-enhanced laser desorption/ionisation (SELDI), enzyme-linked immunosorbent assay (ELISA), surface plasmon resonance (SPR), nano-liquid chromatography-mass spectrometry (nano-LC-MS)

Firstly is the "sequence proteomics" in which amino acid sequences in a given protein are determined. Historically, amino acid compositions were elucidated using Edman sequencing [5]. Briefly, this technique uses a chemical called phenyl isothiocyanate and a mild acid hydrolysis to tag and cleave specifically at the N-terminal of the chosen protein (Fig. 3.3). The amino acids "released" from the whole protein will be identified sequentially using chromatographic instruments such as high-pressure liquid chromatography (HPLC) to build the order of the protein sequences. Up till now, this technique is still considered as one of the most sensitive techniques for protein identification and can sequence down to 0.5–1.0 pmole of proteins [4]. However, this procedure requires a non-complex protein mixture as well as laborious and hence not practical to be used for a larger scale of protein identification. This ultimately requires a high-throughput system such as mass spectrometry (MS) which will be discussed later in this chapter.

The second proteomics area is called "structural proteomics". This study concerns protein structural identity to elucidate its putative function. Structural proteomics can be studied using several approaches including computer-based modelling as well as experimental methods such as protein crystallisation [6], nuclear magnetic resonance (NMR) and electron microscopy [6, 7]. X-ray diffraction of protein crystals is one of the most common techniques in elucidating protein structure in this area.

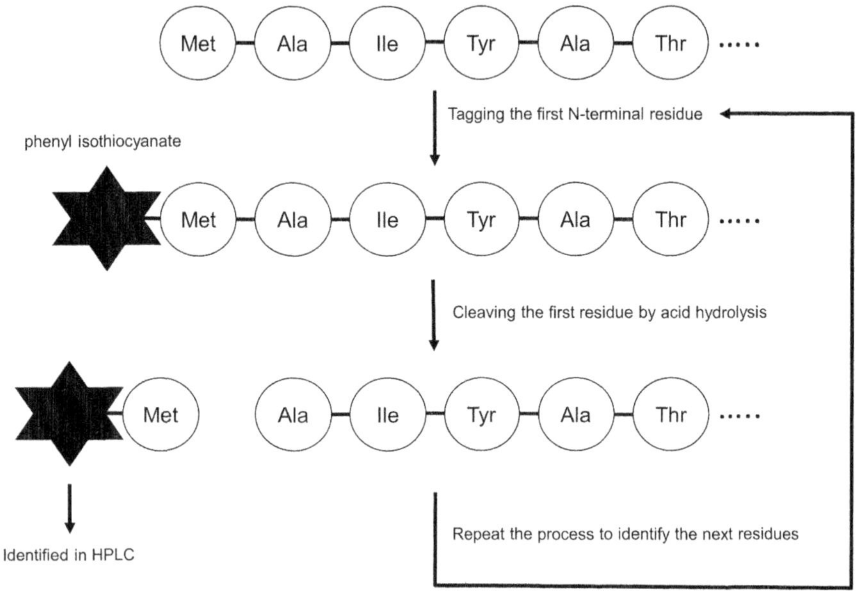

Fig. 3.3 Overall workflow of Edman N-terminal sequencing. The first residue of a protein will be tagged at the N-terminal using phenyl isothiocyanate before cleaved by acid hydrolysis. The identity of the released residue will be determined using analytical instruments such as high-pressure liquid chromatography (HPLC), and the process will be repeated to identify the next residue of the protein, sequentially

Protein functions and activities are studied in the third area of proteomics called "functional and interaction proteomics". Protein function particularly enzymes can be elucidated by examining their reactions in vitro [8]. Other techniques to determine protein function are also available, particularly based on the protein interaction with other proteins, ligands or substrates. Traditionally, techniques such as yeast -one/two-hybrid are a popular tool to elucidate protein-protein/DNA interaction [4]. The introduction of protein microarray experiments has opened a new door for mass characterisation of proteins and protein profiling [9, 10]. Protein post-translational modifications are also able to be detected using certain specific protein arrays [11]. These technologies are based on the interaction of proteins, antibodies and enzymes to other proteins and ligands. One drawback of this approach is it requires known antibodies/enzymes/proteins and hence can be considered as a more targeted proteomics approach.

Last but not least is the "expression proteomics" or also known as "discovery-based proteomics". This approach is useful in elucidating the expression of proteins in a global and untargeted manner. Most proteins in a complex sample can be identified and quantitated to provide an overall protein overview of any experimental samples. This will be highly advantageous for understanding the samples' protein composition as well as finding protein biomarkers. This area of proteomics mainly utilised MS instruments to allow high-throughput analysis of protein/peptide samples. Recently, a proteome map draft of a human has been reported using these instruments which details proteins found in various parts of the body [12, 13]. This suggests that "expression proteomics" is greatly advantageous for a holistic and large-scale study at a system level.

3.1.2 Proteomics in the Context of Systems Biology

As one of the approaches in systems biology, proteomics has been utilised in an integrative approach, combining other omics such as genomics, transcriptomics and metabolomics in an effort to comprehensively understand certain biological questions. Rather than investigating isolated parts of genes/proteins in an organism as what traditional molecular biology have done, a system approach is more useful in characterising the dynamics and structure of a working biological system [14]. This will assist in developing biological models that could be tested upon series of perturbation experiments [15]. Understanding organisms at the proteome level will undeniably contribute to the rationale of the models, considering that proteins are highly modified (due to PTMs) and hence functionally diverse compared to the more static genome.

Due to the nature of systems biology study, not all the different techniques of proteomics can be fully utilised in this area. For example, "sequence proteomics" which is only dedicated for determining protein sequence at a small scale may not be feasible to be used extensively in a larger scale of system studies. On the other hand, "expression proteomics", which can catalogue total quantifiable proteins in any biological system, can be used as the starting platform for a global proteome analysis. Perturbation experiments in any given samples may also be investigated using this approach. One example is using stable isotope labelling by/with amino acids in cell culture (SILAC) which labels specific amino acids with either light or heavy isotopes to investigate the level of protein differences between normal and treated cell lines [16]. Further experimentation can then be employed to characterise any proteins of interest using a more targeted proteomics approach such as in "structural proteomics". Elucidating the protein structure will give an insight to its active sites and how this protein contributes to a given treatment or diseases. Furthermore, protein candidates can be further scrutinised in the "functional and interaction proteomics" approach by finding interacting partners or ligands. Ultimately, these expression, structural and functional protein information can be used to design a workable model for a biological system and hence can be tested in systematic series of perturbation experiments.

3.2 Applications of Proteomics

Proteomics especially the expression proteomics has been applied in various organisms and samples including human cancer and plant crops. These are detailed in the next subsections.

3.2.1 Cancer Proteomics

One of the most studied topics in human is perhaps cancer. Cancer is a complex disease that reflects the genetic as well as protein changes within the cell. Although many effective therapies are present for early detection and diagnosis, cancer remains a major cause of death worldwide, accounting for 8.2 million deaths in 2012 [17]. The most common causes of cancer death are cancers of the lung, liver, stomach, colorectum, breast and oesophagus [18]. In the next two decades, the number of new cancer cases is expected to reach an overwhelming 23.6 million cases [18]. This suggests that more works need to be done to investigate the cause and possible treatment for cancer. Fundamental research particularly at the proteome level will undeniably shed some light into the protein changes that may signal or contribute to the cancer regulation.

Proteomics approaches have been increasingly used for differential analysis of various biological samples from cancer patients, including cell lysates, cell secretome, serum, plasma, tumour tissue and body fluids (Table 3.1). This could lead to a better understanding of the molecular basis of cancer pathogenesis, which can spark the discovery of novel cancer-specific biomarkers [19, 20]. The identification of new cancer biomarkers with predictive value is necessary to allow detection and treatment of cancer when it is still curable [20]. In the different types of cancer, the discovery of biomarkers is supposed to improve one or more of the following critical applications: early diagnosis, prognosis and monitoring of disease progression, its response to therapy as well as its recurrence [21]. Table 3.1 summarises several biomarker discoveries in the lung, pancreatic and gastric cancer, using expression proteomics approach.

3.2.2 Crop Proteomics

Changes in global climate behaviour have resulted in the increase of extreme temperature related phenomena including drought, flood, wind, water erosion and storms [32], which in turn influence soil condition [33]. These changes have already negatively affected the production of staple foods, such as maize, wheat, rice and soybean [32, 34]. Every year, more than 50% of yield loss of major crop plants was estimated worldwide due to abiotic stress such as drought, salinity and extreme temperatures [35]. Moreover, as agriculture land becomes less available, farmers are forced to make use of marginal, low-quality soils, which may contain low levels of nutrients [33].

Plant stress response represents an active process that targets at an establishment of novel homeostasis under altered environmental conditions [36]. Elucidation of the molecular mechanisms underlying the plant response to abiotic stress and the development of stress-tolerant plants have received much attention in recent years. Furthermore, understanding the mechanisms through which plant cells tolerate

Table 3.1 Recent studies on cancer biomarkers using expression proteomics approach

Types	Sample	Potential biomarker	References
Lung cancer	Sera	Protein gene product 9.5 (PGP 9.5)	Brichory et al. [22]
	Cell culture	A disintegrin and metalloprotease-17 or ADAM metallopeptidase domain 17 or tumour necrosis factor-α-converting enzyme (ADAM-17) Osteoprotegerin Pentraxin 3 Follistatin Tumour necrosis factor receptor superfamily 1A	Planque et al. [23]
	Serum	Haptoglobin (HP) Apolipoprotein 4	Okano et al. [24]
	Blood plasma	Apolipoprotein E	Rice et al. [25]
	Cell lines	Heat-shock protein 90-beta (Hsp90-beta) Vimentin (VIM)	Zhang et al. [26]
Pancreatic Cancer	Urine	Lymphatic vessel endothelial hyaluronan receptor (LYVE1) Regenerating islet-derived 1 beta (REG1B) Trefoil factor 1 (TFF1)	Radon et al. [27]
	Plasma cells	Anterior gradient homolog 2 (AGR2) Polymeric immunoglobulin receptor (PIGR) Olfactomedin-4 (OLFM4) Syncollin (SYCN) Collagen alpha-1 (VI) chain (COL6A1)	Makawita et al. [28]
	Serum	Cyclin I Rab GDP dissociation inhibitor ß (GDI2) Haptoglobin precursor Serotransferrin precursor	Sun et al. [29]
Gastric cancer	Tissue	Glucose-regulated protein 78 (GRP78) Glutathione S-transferase pi (GST pi) Apolipoprotein AI (Apo AI) Alpha-1 antitrypsin (A1AT) Gastrokine-1 (GKN-1)	Wu et al. [30]
	Tissue	NSP3 Transgelin (SM22-alpha) Prohibitin Heat-shock 27 kDa protein Protein disulphide isomerase A3 Apolipoprotein AI (ApoAI) Alpha-1 antitrypsin (A1AT)	Ryu et al. [31]

these stresses is essential for the improvement of crop tolerance by genetic engineering or genome editing. Proteomics approaches in particular "expression proteomics" have enabled characterisation of target regulatory proteins and biomarker identification to further comprehend the plant physiology and molecular defence under abiotic stresses.

Drought is one of the major abiotic stresses in crops such as wheat and rice. The proteome of two different wheat varieties with different tolerance to drought, Opata M85 (sensitive) and Nesser (tolerant), were evaluated using abscisic acid (ABA) treatment [37]. Abscisic acid is the key phytohormone produced in response to drought and is involved in coordinating various signalling and metabolic pathways during drought stress. Analysis of their root protein profiles showed that abscisic acid affected the expression level of 805 proteins, and several proteins showed variety-specific regulation by abscisic acid, suggesting their role in drought adaptation [37]. Similarly, physiological analysis of leaf and root protein expression from drought-tolerant wild wheat indicated that abscisic acid level was greatly increased in the drought-treated plants, but the increase was greater and more rapid in the leaves than in the roots [38]. Phosphoproteome analysis of seedling leaves from two bread wheat cultivars (Hanxuan 10 and Ningchun 47) subjected to drought stress also found several important regulators of abscisic acid signalling [39]. To unravel the mechanism behind the maize phosphoenolpyruvate carboxylase (PEPC) gene's capability in improving wheat resistance to drought stress, Qin et al. [40] examined proteome changes under drought conditions of two PEPC-containing transgenic wheat lines and the parental control line (Zhoumai19). The expression of several proteins which related to photosynthesis and cytoskeleton synthesis, and also S-adenosylmethionine synthetase, was induced in transgenic wheat under drought stress, thus demonstrating the efficacy of PEPC in crop improvement.

Proteomics studies have also been performed in rice for drought responses. A number of drought-tolerant and drought-sensitive cultivars of *Oryza sativa* L. ssp. *indica* and *O. sativa* L. *japonica* have been examined [41–44]. Several genes and proteins involved in drought response were identified and characterised. These studies have identified a set of proteins that are drought responsive, including 42 in leaf [44], 22 in rice root [43], 31 in peduncle tissue [42] and 53 in leaf [41]. The understanding of drought responses in rice is critical for designing breeding strategies to develop varieties which are more tolerant to water deficit.

Heat stress is also one of the major stresses affecting crop production, and proteomics has been applied to investigate the molecular response of wheat and rice to such stress. In wheat, proteins related to desiccation and oxidative stress [45, 46], photosynthesis, glycolysis, stress defence, heat shock and ATP production [47] were differentially expressed in the tolerant and sensitive cultivar under heat stress treatment. Meanwhile, heat stress induced the increase of small heat shock protein, β-expansins and lipid transfer proteins in the resistant rice cultivar [48]. This suggests that heat induced protein changes related to stress tolerance and biochemical modifications.

Another abiotic stress, high-salinity condition is also being investigated in various proteomics studies. For example, Xu et al. [49] identified 14 proteins involved in rice seed imbibition during salt stress, in which the majority of these proteins were involved in energy supply and storage protein. Meanwhile, several novel salt stress-responsive proteins, including protein synthesis inhibitor I, photosystem II stability/assembly factor HCF136, trigger factor-like protein and cycloartenol-C24-methyltransferase, were upregulated upon salt stress in rice shoot [50]. Jankangram and colleagues [51] identified ten differential proteins, including gene products involved in photosynthesis, carbon assimilation and the oxidative stress response. They also found that although salinity-sensitive cultivar (Khao Dawk Mali 105) contains elevated transcript level of genes needed for salt tolerance, the post-transcriptional mechanisms controlling protein expression levels were not as efficient as in Pokkali (salinity-tolerant cultivar) [51]. This highlights the importance of studying plant molecular responses at the proteome level, especially during abiotic stresses.

3.3 Proteomics Workflow

Expression proteomics is most useful for cataloguing proteins and finding protein biomarkers in any given samples, especially in the context of systems biology. Hence this subchapter discusses the main experimental design and consideration in the expression proteomics as well as the strategies used to achieve whole proteome analysis.

3.3.1 *Research Design and Consideration*

Expression proteomics in general may consist of five main stages as illustrated in Fig. 3.4. In brief, after identification of the organism of interest, a suitable protein extraction protocol needs to be established before the protein amount is accurately calculated. Certain amounts of proteins are separated followed by identification using a mass spectrometry (MS) instrument. Raw MS data will then be processed using appropriate software to determine the protein identity.

Protein extraction is one of the most critical steps in a proteomics study. This is because high protein yield and clean samples will generate the best proteome coverage. Protocols can vary between organisms [52, 53] and therefore requires thorough literature search and a few rounds of preliminary experiments to determine the best extrac-

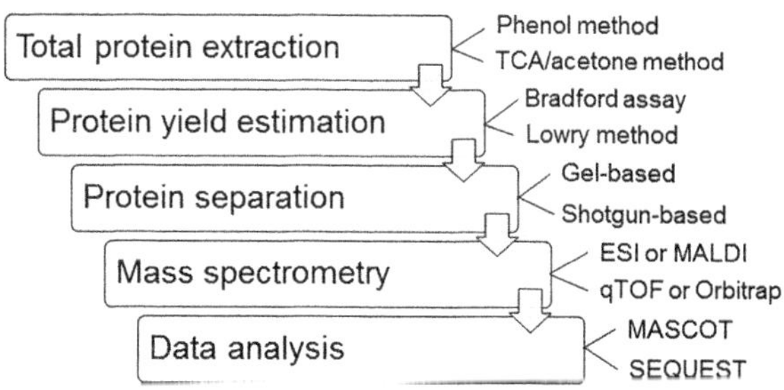

Fig. 3.4 General workflow of an expression proteomics study and examples related to each step (detailed in the text). TCA, trichloroacetic acid; ESI, electrospray ionisation; MALDI, matrix-assisted laser desorption/ionisation; qTOF, quadrupole time-of-flight

tion methods. The two most common methods are phenol and trichloroacetic acid (TCA)/acetone procedures [54, 55]. Determining protein levels in extracted samples is also crucial to evaluate differences between extraction methods and for consistent loading into gels and columns. Protein concentration is often estimated by interpolation of samples' absorbance on a protein standard curve, normally constructed using different concentrations of protein standards such as bovine serum albumin (BSA). A few methods are widely used, namely, Bradford assay and Lowry method [56, 57].

Total protein extracts then need to be first separated to simplify the protein mixture before protein identification can be done. There are two different strategies for this, namely, gel-based and shotgun-based approaches [53, 58, 59]. Gel-based platform employs a two-dimensional gel electrophoresis (2DGE) system which essentially separates the proteins based on their isoelectric point (pI) and molecular weight [60, 61]. Meanwhile, shotgun-based proteomics rely on the resolving power of chromatographic techniques that can separate biomolecules using ion exchange and reversed-phase columns, among others [4]. There are several differences between gel-based and shotgun-based techniques [4, 52]. Firstly, 2DGE is often regarded more laborious as a gel medium is needed to resolve protein spots, whereas chromatographic techniques can often be an on-line procedure without the need of any resolving gel medium beforehand. Secondly, gel-based proteomics require visual inspection and densitometer to quantify protein spot differences between samples [62]. Whereas in shotgun-based approach, quantitative measurement of peptides (corresponding to the proteins) can be achieved through isotope labelling (such as SILAC) or spectral counting for labelled-free approach [63, 64]. Thirdly, protein digestion is required after 2DGE separation for gel-based approach, yet protein mixtures are digested even before the chromatographic run in the shotgun-based workflow. Finally, unlike 2DGE which is not easily automated (although robotic arms for spot picking exist), the chromatography columns often can be coupled with downstream mass spectrometry analysis for peptide identification.

Once proteins are successfully separated and digested, mass spectrometry (MS) will be used to elucidate the molecular mass of these peptides. MS instruments are consisted of three sequentially ordered parts: [1] ion source [2], mass analyser and [3] detector [64–66]. This has been illustrated in Fig. 3.5. Before any peptide samples are able to be measured, they first need to be ionised using the ion source, which can often be either an electrospray ionisation (ESI) or matrix-assisted laser desorption/ionisation (MALDI). Liquid peptide samples separated using liquid chromatography are well suited for ESI and hence a method of choice for shotgun proteomics, whereas single/a few protein samples isolated from gels (SDS-PAGE or 2DGE) are commonly using MALDI for peptide identification. The next part of a MS instrument is the mass analyser which essentially separates the peptide ions before fragmenting them. This generates a profile of peptide ions differing in masses and charges which are denoted as *m/z* values. Most commonly used mass analysers in proteomics studies are quadrupole time-of-flight (qTOF) and Orbitrap, owing to their sensitivity, accuracy and speed [65]. The resulting peptide ions are then measured in the third part of a MS called detector which will supply the data to designated workstations. This generated raw data then need to be thoroughly analysed using appropriate platforms/software.

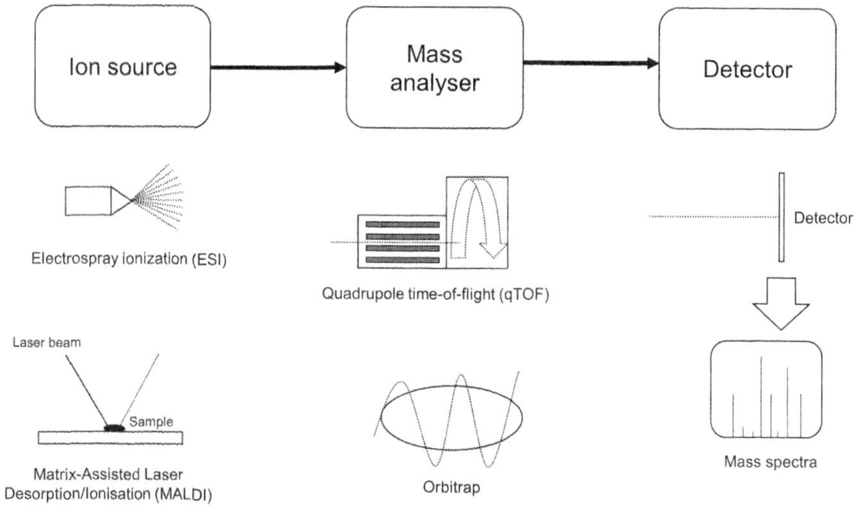

Fig. 3.5 Key components of mass spectrometry (MS) with representative figures of techniques/instruments used in each component. It consists of an ion source which can be either an electrospray ionisation (ESI) or matrix-assisted laser desorption/ionisation (MALDI) type to ionise sample peptides (dotted lines). Once ionised, peptides will be separated based on mass and charge in mass analysers which include quadrupole time-of-flight (qTOF) and Orbitrap. The detector then computes the peptide information (mass spectra) from the mass analysers to be used in subsequent data analysis

There are a number of software available for MS data analysis. Commonly used proteomics software are Mascot (www.matrixscience.com) and SEQUEST (fields.scripps.edu/sequest/) [63]. Other software that are available for proteomics data analysis are listed in Twyman [4] and Rose et al. [53]. These software are mainly differed in their algorithms for protein identification but essentially consist of six main steps as detailed in Twyman [4]. The first two steps involve MS spectral data collection and processing. Then protein sequences from specific databases of organism of interest are theoretically digested using enzymes similarly used for the protein samples earlier. The next steps are processing the theoretical spectra and matching them with the processed MS spectral data. Finally, statistical analysis is required to measure how good is the match between the MS and theoretical spectra for accurate identification of the protein ID. These six main steps are the basis of peptide mass fingerprinting (PMF). De novo protein sequence identification using software such as Peaks and PepNovo is also another alternative [67]. For publication, minimal information about a proteomics experiment (MIAPE, www.psidev.info/node/91) has been established and required in top proteomics journals such as *Proteomics* as well as *Molecular and Cellular Proteomics* [55, 68]. Hence, these requirements, such as depositing data into a public domain [69], need to be followed to ensure successful manuscript revision and publication.

3.3.2 Different Strategies for Expression Proteomics

Both gel-based and shotgun-based approaches (Fig. 3.6) have various improvised strategies to quantify proteins from different samples, often simultaneously. For example, 2D-DIGE (differential gel electrophoresis) method labels different samples with different florescent dyes before 2DGE is performed [62]. Protein abundance will be quantified according to its protein spot signals measured using an imager [62]. On the other hand, shotgun-based approach has also employed similar approach where proteins are labelled to be quantified. Different variations of this approach were reported such as isotope-coded affinity tagging (ICAT), isobaric tags for relative and absolute quantitation (iTRAQ) and SILAC [52]. These techniques used isotopes to label proteins/peptides, and differences between peptides from different samples will be quantitated using appropriate MS analysis. More recently, techniques such as multidimensional protein identification technology

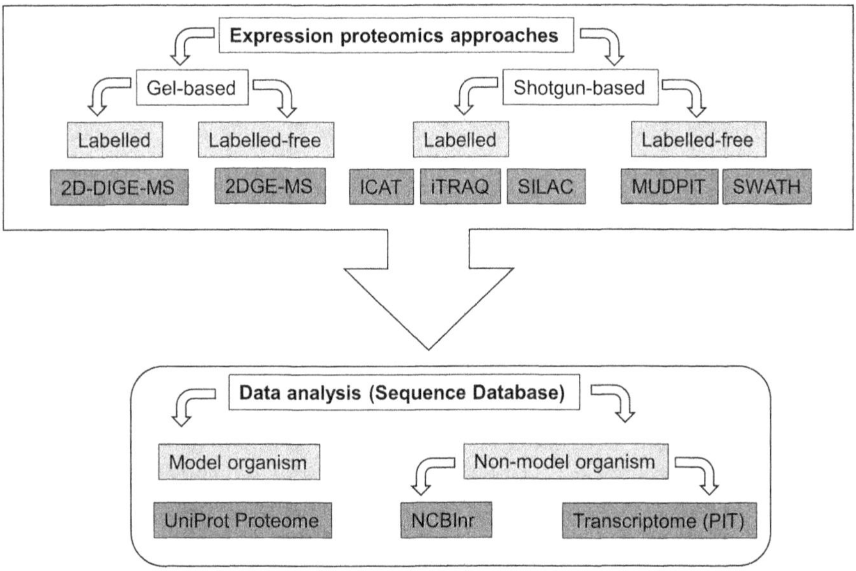

Fig. 3.6 Expression proteomics can be divided either into gel-based or shotgun-based approaches. For labelled and labelled-free gel-based approaches, two-dimensional-differential gel electrophoresis (2D-DIGE) method and 2D gel electrophoresis (2DGE) are, respectively, used which then coupled with mass spectrometry (MS) for protein identification. For labelled shotgun-based approach, isotope-coded affinity tagging (ICAT), isobaric tags for relative and absolute quantitation (iTRAQ) and stable isotope labelling by/with amino acids in cell culture (SILAC) are commonly used, whereas labelled-free utilises either multidimensional protein identification technology (MUDPIT) or sequential window acquisition of all theoretical mass spectra (SWATH-MS). For data analysis, model organism can use its available proteome sequence database from the UniProt website (https://www.uniprot.org/), whereas non-model organism may use either NCBI nonredundant database (NCBInr) or corresponding transcriptome database in a strategy called proteomics informed by transcriptomics (PIT)

(MUDPIT) coupled with Orbitrap [52, 70] and sequential window acquisition of all theoretical mass spectra (SWATH-MS) analysis using TripleTOF technology [71] have been developed and used successively to quantify large number of proteins without labelling. All in all, these various different strategies allow high-throughput and sensitive approaches for quantitative proteomics and have certainly propelled this research area.

However, proteomics approach has often been hindered by the lack of reference database (Fig. 3.6). For model organisms, their complete proteome sequence database can be easily obtained through UniProt database (https://www.uniprot.org/). The availability of a specific and complete protein sequence database is crucial for peptide mass spectra to be analysed and accurately predicted using PMF [4, 68]. This would help protein identification from spectral peptides and hence determine their possible function and biological relevance. PMF without a complete genome/protein database can be a daunting task for any bioinformatics tools as a number of non-significant hits can be generated even when using NCBI nonredundant database (NCBInr) [68]. Therefore, proteomics informed by transcriptomics (PIT) strategy has been introduced for the proteomics analysis of non-model organisms [72, 73]. Using the sequence data information obtained through transcriptomic analysis (detailed in Chap. 2), the protein profile of a given organism can be correctly deduced [72]. Given that a number of herbs and exotic plants for proteomics, which genome sequences are largely unavailable, PIT is the best strategy to be opt for in the future.

3.4 Proteomics and Enzymatic Studies of Kesum

3.4.1 Kesum: A Proteomics Case Study

Proteomics approach has been used in several plant species including model plant Arabidopsis [74] and fruit crops, such as tomato [75, 76] and capsicum [77]. However, proteomics studies in non-model organisms are often hindered by the lack of reference sequence database. While it is still possible to use general public sequence database particularly from NCBI, this often leads to lesser number of identified proteins due to low match hits. Nonetheless, proteomics studies in herbal plant species have been performed in a few species including *Gastrodia elata* orchid [78], *Zanthoxylum nitidum* [79] and *Pueraria radix* [80]. Other non-model species such as *Persicaria minor* Huds. or locally known as kesum, pitcher plant (*Nepenthes* sp.), mahogany (*Swietenia macrophylla* King) seeds and mangosteen (*Garcinia mangostana* L.) fruit are also of interest in this tropical region and currently being investigated using proteomics approach.

Kesum in particular has been used in many traditional cuisine across Southeast Asian countries including Malaysia because of its pungent smell [81]. Moreover, several studies have characterised that the plants contain certain medicinal compounds that exhibit antibacterial and anticancer properties [81]. It is also being used as antidandruff as well as aromatherapy products [82, 83]. Some of the many compounds in kesum can be classified as terpenoids and aliphatic aldehyde compounds which contribute to the smell and taste of this herbal plant [84–86]. In order to investigate the biosynthetic pathways of these compounds, a proteomics analysis employing a shotgun proteomics has been performed in this species. Several proteins have been identified to be differentially expressed upon methyl jasmonate treatment [87] which may be responsible for the induction of different volatile compounds during stress conditions [85, 70]. This herbal proteomics study [87] employed PIT to increase the number of peptide match hits and ultimately assisting the protein identification.

3.4.2 Enzymatic Studies from Kesum

Enzymes of secondary metabolites biosynthetic pathway are attractive targets for development of potential antimicrobial, anticancer drugs and insect-resistant crop plants by deployment of transgenic plant. However, as the numbers of known genes are growing, the elucidation of their functions remains a major bottleneck and lag behind the sequencing capability [88]. Therefore, more studies need to be done to characterise enzymes and their substrates as well as products to understand their activities and mechanistic properties.

Functional characterisation of selected proteins from kesum has also been performed using enzymatic analysis. The research aimed at identifying substrates and products of oxidoreductase enzymes, which are involved in the biosynthetic pathway of monoterpenes and sesquiterpenes in kesum [89–92]. Several buffer compositions containing reducing agents, osmotic reagents, protease inhibitors and phenolic absorbent were employed to select the most suitable extraction buffers for the extraction of selected enzymes [89–92]. These enzymes including terpene alcohol dehydrogenases, terpene aldehyde dehydrogenases and terpene synthases were identified in the cell-free extract of kesum leaves.

References

1. Humphery-Smith I (2015) The 20th anniversary of proteomics and some of its origins. Proteomics 15:1773–1776
2. Banks RE, Dunn MJ, Hochstrasser DF, Sanchez J-C, Blackstock W, Pappin DJ, Selby PJ (2000) Proteomics: new perspectives, new biomedical opportunities. Lancet 356:1749–1756
3. Khoury GA, Baliban RC, Floudas CA (2011) Proteome-wide post-translational modification statistics: frequency analysis and curation of the swiss-prot database. Sci Rep 1:1–5
4. Twyman RM (2013) Principles of proteomics. Garland Science, Abingdon
5. Edman P, Begg G (1967) A Protein Sequenator. Eur J Biochem 1:80–91

6. Sali A, Glaeser R, Earnest T, Baumeister W (2003) From words to literature in structural proteomics. Nature 422:216–225
7. Woolfson M (2018) The development of structural x-ray crystallography. Phys Scr 93:1–32
8. Bisswanger H (2014) Enzyme assays. Perspect Sci 1:41–55
9. LaBaer J, Ramachandran N (2005) Protein microarrays as tools for functional proteomics. Curr Opin Chem Biol 9:14–19
10. Reymond Sutandy FX, Qian J, Chen C-S, Zhu H (2013) Overview of protein microarrays. Curr Protoc Protein Sci 72:1–21
11. Lueong SS, Hoheisel JD, Alhamdani MSS (2014) Protein microarrays as tools for functional proteomics: achievements, promises and challenges. J Proteomics Bioinform 7:1–10
12. Kim M-S, Pinto SM, Getnet D, Nirujogi RS, Manda SS, Chaerkady R, Madugundu AK, Kelkar DS, Isserlin R, Jain S, Thomas JK, Muthusamy B, Leal-Rojas P, Kumar P, Sahasrabuddhe NA, Balakrishnan L, Advani J, George B, Renuse S, Selvan LDN, Patil AH, Nanjappa V, Radhakrishnan A, Prasad S, Subbannayya T, Raju R, Kumar M, Sreenivasamurthy SK, Marimuthu A, Sathe GJ, Chavan S, Datta KK, Subbannayya Y, Sahu A, Yelamanchi SD, Jayaram S, Rajagopalan P, Sharma J, Murthy KR, Syed N, Goel R, Khan AA, Ahmad S, Dey G, Mudgal K, Chatterjee A, Huang T-C, Zhong J, Wu X, Shaw PG, Freed D, Zahari MS, Mukherjee KK, Shankar S, Mahadevan A, Lam H, Mitchell CJ, Shankar SK, Satishchandra P, Schroeder JT, Sirdeshmukh R, Maitra A, Leach SD, Drake CG, Halushka MK, Prasad TSK, Hruban RH, Kerr CL, Bader GD, Iacobuzio-Donahue CA, Gowda H, Pandey A (2014) A draft map of the human proteome. Nature 509:575–581
13. Wilhelm M, Schlegl J, Hahne H, Gholami AM, Lieberenz M, Savitski MM, Ziegler E, Butzmann L, Gessulat S, Marx H, Mathieson T, Lemeer S, Schnatbaum K, Reimer U, Wenschuh H, Mollenhauer M, Slotta-Huspenina J, Boese J-H, Bantscheff M, Gerstmair A, Faerber F, Kuster B (2014) Mass-spectrometry-based draft of the human proteome. Nature 509:582–587
14. Kitano H (2002) Systems biology: a brief overview. Science 295:1662–1664
15. Ideker T, Galitski T, Hood L (2001) A new approach to decoding life: systems biology. Annu Rev Genomics Hum Genet 2:343–372
16. Ong S-E, Blagoev B, Kratchmarova I, Kristensen DB, Steen H, Pandey A, Mann M (2002) Stable isotope labeling by amino acids in cell culture, SILAC, as a simple and accurate approach to expression proteomics. Mol Cell Proteomics 1:376–386
17. www.cruk.org/cancerstats (2014)
18. www.cancer.gov (2014)
19. Savino R, Paduano S, Preianò M, Terracciano R (2012) The proteomics big challenge for biomarkers and new drug-targets discovery. Int J Mol Sci 13:13926–13948
20. Maurya P, Meleady P, Dowling P, Clynes M (2007) Proteomic approaches for serum biomarker discovery in cancer. Anticancer Res 27:1247–1255
21. Sallam RM (2015) Proteomics in cancer biomarkers discovery: challenges and applications. Dis Markers 2015:1–12
22. Brichory F, Beer D, LeNaour F, Giordano T, Hanash S (2001) Proteomics-based identification of protein gene product 9.5 as a tumor antigen that induces a humoral immune response in lung cancer. Cancer Res 61:7908–7912
23. Planque C, Kulasingam V, Smith CR, Reckamp K, Goodglick L, Diamandis EP (2009) Identification of five candidate lung cancer biomarkers by proteomics analysis of conditioned media of four lung cancer cell lines. Mol Cell Proteomics 8:2746–2758
24. Okano T, Seike M, Kuribayashi H, Soeno C, Ishii T, Kida K, Gemma A (2016) Identification of haptoglobin peptide as a novel serum biomarker for lung squamous cell carcinoma by serum proteome and peptidome profiling. Int J Oncol 48:945–952
25. Rice SJ, Liu X, Miller B, Joshi M, Zhu J, Caruso C, Gilbert C, Toth J, Reed M, Rassaei N (2015) Proteomic profiling of human plasma identifies apolipoprotein E as being associated with smoking and a marker for squamous metaplasia of the lung. Proteomics 15:3267–3277

26. Zhang W, Li Y, Yang S, Li W, Ming Z, Zhang Y, Hou Y, Niu Z, Rong B, Zhang X (2013) Differential mitochondrial proteome analysis of human lung adenocarcinoma and normal bronchial epithelium cell lines using quantitative mass spectrometry. Thorac Cancer 4:373–379
27. Radon TP, Massat NJ, Jones R, Alrawashdeh W, Dumartin L, Ennis D, Duffy SW, Kocher HM, Pereira SP, Guarner L (2015) Identification of a three-biomarker panel in urine for early detection of pancreatic adenocarcinoma. Clin Cancer Res 21:3512–3521
28. Makawita S, Smith C, Batruch I, Zheng Y, Rückert F, Grützmann R, Pilarsky C, Gallinger S, Diamandis EP (2011) Integrated proteomic profiling of cell line conditioned media and pancreatic juice for the identification of pancreatic cancer biomarkers. Mol Cell Proteomics 10:1–21
29. Sun Z-L, Zhu Y, Wang F-Q, Chen R, Peng T, Fan Z-N, Xu Z-K, Miao Y (2007) Serum proteomic-based analysis of pancreatic carcinoma for the identification of potential cancer biomarkers. Biochimica Biophys Acta, Proteins Proteomics 1774:764–771
30. Wu J-Y, Cheng C-C, Wang J-Y, Wu D-C, Hsieh J-S, Lee S-C, Wang W-M (2014) Discovery of tumor markers for gastric cancer by proteomics. PLOS ONE 9:e84158
31. Ryu J-W, Kim H-J, Lee Y-S, Myong N-H, Hwang C-H, Lee G-S, Yom H-C (2003) The proteomics approach to find biomarkers in gastric cancer. J Korean Med Sci 18:505–509
32. FAO (2015) Climate change and food security: risks and responses. Food and Agriculture Organization of the United Nations (FAO). www.fao.org
33. Barkla BJ (2016) Identification of abiotic stress protein biomarkers by proteomic screening of crop cultivar diversity. Proteomes 4:26
34. Kang Y, Khan S, Ma X (2009) Climate change impacts on crop yield, crop water productivity and food security – a review. Prog Nat Sci 19:1665–1674
35. Qin F, Shinozaki K, Yamaguchi-Shinozaki K (2011) Achievements and challenges in understanding plant abiotic stress responses and tolerance. Plant Cell Physiol 52:1569–1582
36. Kosová K, Vítámvás P, Urban MO, Klíma M, Roy A, Prášil IT (2015) Biological networks underlying abiotic stress tolerance in temperate crops—a proteomic perspective. Int J Mol Sci 16:20913–20942
37. Alvarez S, Roy Choudhury S, Pandey S (2014) Comparative quantitative proteomics analysis of the ABA response of roots of drought-sensitive and drought-tolerant wheat varieties identifies proteomic signatures of drought adaptability. J Proteome Res 13:1688–1701
38. Liu H, Sultan MARF, li Liu X, Zhang J, Yu F, Xian Zhao H (2015) Physiological and comparative proteomic analysis reveals different drought responses in roots and leaves of drought-tolerant wild wheat (*Triticum boeoticum*). PLOS ONE 10:e0121852
39. Zhang M, Lv D, Ge P, Bian Y, Chen G, Zhu G, Li X, Yan Y (2014) Phosphoproteome analysis reveals new drought response and defense mechanisms of seedling leaves in bread wheat (*Triticum aestivum* L.). J Proteome 109:290–308
40. Qin N, Xu W, Hu L, Li Y, Wang H, Qi X, Fang Y, Hua X (2016) Drought tolerance and proteomics studies of transgenic wheat containing the maize C4 phosphoenolpyruvate carboxylase (PEPC) gene. Protoplasma 253:1503–1512
41. Maksup S, Roytrakul S, Supaibulwatana K (2014) Physiological and comparative proteomic analyses of Thai jasmine rice and two check cultivars in response to drought stress. J Plant Interact 9:43–55
42. Muthurajan R, Shobbar Z-S, Jagadish S, Bruskiewich R, Ismail A, Leung H, Bennett J (2011) Physiological and proteomic responses of rice peduncles to drought stress. Mol Biotechnol 48:173–182
43. Rabello AR, Guimarães CM, Rangel PH, da Silva FR, Seixas D, de Souza E, Brasileiro AC, Spehar CR, Ferreira ME, Mehta Â (2008) Identification of drought-responsive genes in roots of upland rice (*Oryza sativa* L.). BMC Genomics 9:485
44. Salekdeh GH, Siopongco J, Wade LJ, Ghareyazie B, Bennett J (2002) Proteomic analysis of rice leaves during drought stress and recovery. Proteomics 2:1131–1145
45. Kumar RR, Singh GP, Goswami S, Pathak H, Rai RD (2014) Proteome analysis of wheat (*Triticum aestivum*) for the identification of differentially expressed heat-responsive proteins. Aust J Crop Sci 8:973

46. Laino P, Shelton D, Finnie C, De Leonardis AM, Mastrangelo AM, Svensson B, Lafiandra D, Masci S (2010) Comparative proteome analysis of metabolic proteins from seeds of durum wheat (cv. Svevo) subjected to heat stress. Proteomics 10:2359–2368
47. Wang X, Dinler BS, Vignjevic M, Jacobsen S, Wollenweber B (2015) Physiological and proteome studies of responses to heat stress during grain filling in contrasting wheat cultivars. Plant Sci 230:33–50
48. Mu Q, Zhang W, Zhang Y, Yan H, Liu K, Matsui T, Tian X, Yang P (2017) iTRAQ-based quantitative proteomics analysis on rice anther responding to high temperature. Int J Mol Sci 18:1811
49. Xu E, Chen M, He H, Zhan C, Cheng Y, Zhang H, Wang Z (2017) Proteomic analysis reveals proteins involved in seed imbibition under salt stress in rice. Front Plant Sci 7:2006
50. Li X-J, Yang M-F, Zhu Y, Liang Y, Shen S-H (2011) Proteomic analysis of salt stress responses in rice shoot. J Plant Biol 54:384
51. Jankangram W, Thammasirirak S, Jones MG, Hartwell J, Theerakulpisut P (2011) Proteomic and transcriptomic analysis reveals evidence for the basis of salt sensitivity in Thai jasmine rice (*Oryza sativa* L. cv. KDML 105). Afr J Biotechnol 10:16157–16166
52. Jorrín-Novo JV, Pascual J, Sánchez-Lucas R, Romero-Rodríguez MC, Rodríguez-Ortega MJ, Lenz C, Valledor L (2015) Fourteen years of plant proteomics reflected in Proteomics: moving from model species and 2DE-based approaches to orphan species and gel-free platforms. Proteomics 15:1089–1112
53. Rose JKC, Bashir S, Giovannoni JJ, Jahn MM, Saravanan RS (2004) Tackling the plant proteome: practical approaches, hurdles and experimental tools. Plant J 39:715–733
54. Carpentier SC, Witters E, Laukens K, Deckers P, Swennen R, Panis B (2005) Preparation of protein extracts from recalcitrant plant tissues: an evaluation of different methods for two-dimensional gel electrophoresis analysis. Proteomics 5:2497–2507
55. Jorrín-Novo JV, Maldonado AM, Echevarría-Zomeño S, Valledor L, Castillejo MA, Curto M, Valero J, Sghaier B, Donoso G, Redondo I (2009) Plant proteomics update (2007–2008): second-generation proteomic techniques, an appropriate experimental design, and data analysis to fulfill MIAPE standards, increase plant proteome coverage and expand biological knowledge. J Proteome 72:285–314
56. Redmile-Gordon MA, Armenise E, White RP, Hirsch PR, Goulding KWT (2013) A comparison of two colorimetric assays, based upon Lowry and Bradford techniques, to estimate total protein in soil extracts. Soil Biol Biochem 67:166–173
57. Okutucu B, Dınçer A, Habib Ö, Zıhnıoglu F (2007) Comparison of five methods for determination of total plasma protein concentration. J Biochem Biophys Methods 70:709–711
58. Martínez-Esteso MJ, Martínez-Márquez A, Sellés-Marchart S, Morante-Carriel JA, Bru-Martínez R (2015) The role of proteomics in progressing insights into plant secondary metabolism. Front Plant Sci 6:504
59. Finoulst I, Pinkse M, Van Dongen W, Verhaert P (2011) Sample preparation techniques for the untargeted LC-MS-based discovery of peptides in complex biological matrices. Biomed Res Int 2011:245291
60. Cañas B, Piñeiro C, Calvo E, López-Ferrer D, Gallardo JM (2007) Trends in sample preparation for classical and second generation proteomics. J Chromatogr A 1153:235–258
61. Neverova I, Van Eyk JE (2005) Role of chromatographic techniques in proteomic analysis. J Chromatogr B 815:51–63
62. Issaq HJ, Veenstra TD (2008) Two-dimensional polyacrylamide gel electrophoresis (2D-PAGE): advances and perspectives. BioTechniques 44:697
63. Old WM, Meyer-Arendt K, Aveline-Wolf L, Pierce KG, Mendoza A, Sevinsky JR, Resing KA, Ahn NG (2005) Comparison of label-free methods for quantifying human proteins by shotgun proteomics. Mol Cell Proteomics 4:1487–1502
64. Zhu W, Smith JW, Huang C-M (2009) Mass spectrometry-based label-free quantitative proteomics. Biomed Res Int 2010:840518

65. Domon B, Aebersold R (2010) Options and considerations when selecting a quantitative proteomics strategy. Nat Biotechnol 28:710–721
66. Lin D, Tabb DL, Yates JR (2003) Large-scale protein identification using mass spectrometry. Biochimica Biophys Acta, Proteins Proteomics 1646:1–10
67. Romero-Rodríguez MC, Pascual J, Valledor L, Jorrín-Novo J (2014) Improving the quality of protein identification in non-model species. Characterization of *Quercus ilex* seed and *Pinus radiata* needle proteomes by using SEQUEST and custom databases. J Proteome 105:85–91
68. Jorrín-Novo JV (2015) Scientific standards and MIAPEs in plant proteomics research and publications. Front Plant Sci 6:473
69. Taylor CF, Paton NW, Lilley KS, Binz P-A, Julian RK, Jones AR, Zhu W, Apweiler R, Aebersold R, Deutsch EW (2007) The minimum information about a proteomics experiment (MIAPE). Nat Biotechnol 25:887–893
70. Webb KJ, Xu T, Park SK, Yates JR (2013) Modified MuDPIT separation identified 4488 proteins in a system-wide analysis of quiescence in yeast. J Proteome Res 12:2177–2184
71. Rosenberger G, Koh CC, Guo T, Röst HL, Kouvonen P, Collins BC, Heusel M, Liu Y, Caron E, Vichalkovski A, Faini M, Schubert OT, Faridi P, Ebhardt HA, Matondo M, Lam H, Bader SL, Campbell DS, Deutsch EW, Moritz RL, Tate S, Aebersold R (2014) A repository of assays to quantify 10,000 human proteins by SWATH-MS. Sci Data 1:1–15
72. Evans VC, Barker G, Heesom KJ, Fan J, Bessant C, Matthews DA (2012) *De novo* derivation of proteomes from transcriptomes for transcript and protein identification. Nat Methods 9:1207–1211
73. Mudenda L, Pierlé SA, Turse JE, Scoles GA, Purvine SO, Nicora CD, Clauss TRW, Ueti MW, Brown WC, Brayton KA (2014) Proteomics informed by transcriptomics identifies novel secreted proteins in *Dermacentor andersoni* saliva. Int J Parasitol 44:1029–1037
74. Baerenfaller K, Grossmann J, Grobei MA, Hull R, Hirsch-Hoffmann M, Yalovsky S, Zimmermann P, Grossniklaus U, Gruissem W, Baginsky S (2008) Genome-scale proteomics reveals *Arabidopsis thaliana* gene models and proteome dynamics. Science 320:938–941
75. Faurobert M, Mihr C, Bertin N, Pawlowski T, Negroni L, Sommerer N, Causse M (2007) Major proteome variations associated with cherry tomato pericarp development and ripening. Plant Physiol 143:1327–1346
76. Rocco M, D'Ambrosio C, Arena S, Faurobert M, Scaloni A, Marra M (2006) Proteomic analysis of tomato fruits from two ecotypes during ripening. Proteomics 6:3781–3791
77. Aizat WM, Able JA, Stangoulis JCR, Able AJ (2013) Proteomic analysis during capsicum ripening reveals differential expression of ACC oxidase isoform 4 and other candidates. Funct Plant Biol 40:1115–1128
78. Zeng X, Li Y, Ling H, Chen J, Guo S (2018) Revealing proteins associated with symbiotic germination of *Gastrodia elata* by proteomic analysis. Bot Stud 59:8
79. Lu Q, Zhang ZS, Zhan RT, He R (2018) Proteomic analysis of *Zanthoxylum nitidum* seeds dormancy release: influence of stratification and gibberellin. Ind Crop Prod 122:7–15
80. Kim Y, Chung WS, Jang HJ (2018) Proteins isolated of *Pueraria radix* possible to cause allergenic react with immunoglobulin E in human sera. Mol Cell Toxicol 14:233–239
81. Vikram P, Chiruvella KK, Ripain IHA, Arifullah M (2014) A recent review on phytochemical constituents and medicinal properties of kesum (*Polygonum minus* Huds.). Asian Pac J Trop Biomed 4:430–435
82. Bunawan H, Talip N, Noor NM (2011) Foliar anatomy and micromorphology of *Polygonum minus* Huds. And their taxonomic implications. Vascular 5:5–10
83. Azlim Almey A, Ahmed Jalal Khan C, Syed Zahir I, Mustapha Suleiman K, Aisyah M, Kamarul Rahim K (2010) Total phenolic content and primary antioxidant activity of methanolic and ethanolic extracts of aromatic plants' leaves. Int Food Res J 17:1077–1084
84. Baharum SN, Bunawan H, Ghani M a A, Mustapha WAW, Noor NM (2010) Analysis of the chemical composition of the essential oil of *Polygonum minus* Huds. Using two-dimensional gas chromatography-time-of-flight mass spectrometry (GC-TOF MS). Molecules 15:7006–7015

85. Khairudin K, Sukiran NA, Goh H-H, Baharum SN, Noor NM (2014) Direct discrimination of different plant populations and study on temperature effects by Fourier transform infrared spectroscopy. Metabolomics 10:203–211
86. Goh HH, Khairudin K, Sukiran NA, Baharum SN, Normah M (2015) Metabolite profiling reveals temperature effects on the VOCs and flavonoids of different plant populations. Plant Biol 18:130–139
87. Aizat WM, Ibrahim S, Rahnamaie-Tajadod R, Loke K-K, Goh H-H, Noor NM (2018) Extensive mass spectrometry proteomics data of *Persicaria minor* herb upon methyl jasmonate treatment. Data Brief 16:1091–1094
88. Amin SR, Erdin S, Ward RM, Lua RC, Lichtarge O (2013) Prediction and experimental validation of enzyme substrate specificity in protein structures. Proc Natl Acad Sci 110:E4195–E4202
89. Hassan M, Maarof ND, Ali ZM, Noor NM, Othman R, Mori N (2012) Monoterpene alcohol metabolism: identification, purification, and characterization of two geraniol dehydrogenase isoenzymes from *Polygonum minus* leaves. Biosci Biotechnol Biochem 76:1463–1470
90. Ahmad-Sohdi NAS, Seman-Kamarulzaman A-F, Mohamed-Hussein Z-A, Hassan M (2015) Purification and characterization of a novel NAD (P)+-farnesol dehydrogenase from *Polygonum minus* leaves. PLOS ONE 10:e0143310
91. Seman-Kamarulzaman A-F, Mohamed-Hussein Z-A, Ng CL, Hassan M (2016) Novel NAD+-farnesal dehydrogenase from *Polygonum minus* leaves. Purification and characterization of enzyme in juvenile hormone III biosynthetic pathway in plant. PLOS ONE 11:e0161707
92. Nik-Abdul-Ghani N-R, Mohamed-Hussein Z-A, Hassan M (2017) Citral dehydrogenase involved in geraniol oxidation pathway: purification, characterization and kinetic studies from *Persicaria minor* (*Polygonum minus* Huds.). J Plant Biochem Biotechnol 27:1–12

Chapter 4
Metabolomics in Systems Biology

Syarul Nataqain Baharum and Kamalrul Azlan Azizan

Abstract Over the last decade, metabolomics has continued to grow rapidly and is considered a dynamic technology in envisaging and elucidating complex phenotypes in systems biology area. The advantage of metabolomics compared to other omics technologies such as transcriptomics and proteomics is that these later omics only consider the intermediate steps in the central dogma pathway (mRNA and protein expression). Meanwhile, metabolomics reveals the downstream products of gene and expression of proteins. The most frequently used tools are nuclear magnetic resonance (NMR) spectroscopy and mass spectrometry (MS). Some of the common MS-based analyses are gas chromatography-mass spectrometry (GC-MS) and liquid chromatography-mass spectrometry (LC-MS). These high-throughput instruments play an extremely crucial role in discovery metabolomics to generate data needed for further analysis. In this chapter, the concept of metabolomics in the context of systems biology is discussed and provides examples of its application in human disease studies, plant responses towards stress and abiotic resistance and also microbial metabolomics for biotechnology applications. Lastly, a few case studies of metabolomics analysis are also presented, for example, investigation of an aromatic herbal plant, *Persicaria minor* metabolome and microbial metabolomics for metabolic engineering applications.

Keywords Biomarker discovery · Mass spectrometry · Metabolite profiling · Metabolic pathway · Metabonomics · Phytochemical analysis

---

S. N. Baharum (✉) · K. A. Azizan
Institute of Systems Biology, Universiti Kebangsaan Malaysia (UKM), Bangi, Malaysia
e-mail: nataqain@ukm.edu.my; kamalrulazlan@ukm.edu.my

W. M. Aizat et al. (eds.), *Omics Applications for Systems Biology*,
Advances in Experimental Medicine and Biology 1102,
https://doi.org/10.1007/978-3-319-98758-3_4

4.1 Introduction

Metabolomics is an emerging technology that concerns with the understanding of living organisms at the metabolic level. As a definition, the metabolome represents the whole metabolites within a biological system [1]. It represents the crucial phenotype of cells inferred by the perturbation of gene expression and protein functions, due to the environmental effects or mutations. Gene expression and protein function also can be influenced by metabolome changes. Consequently, metabolomics plays an important role in cellular system and gene functions.

4.1.1 Different Aspects of Metabolomics

Specifically, the study of small molecule metabolite level is defined as metabolomics which is aimed at the measurement of metabolites present in a cell or tissue under a particular set of conditions. Metabolomics is considered an "omics" for the small molecules that provides information on the metabolome set at the individual level. Metabolomics is known by several names such as metabolic profiling, targeted/untargeted metabolic profiling and metabonomics [2]. Nevertheless, these approaches involved the analysis of metabolite levels from plant/microbe extracts, biofluids and living tissues [3]. While metabolomics is well associated with metabolite analysis of plant/microbes, the term metabonomics is usually applied to studies of human. Analysis of biofluids such as urine and plasma, nutrition, responses to drugs or diseases are some of the examples that are being tackled by metabonomics [4].

4.1.2 Metabolomics in the Context of Systems Biology

According to Weckwerth et al. [5], the principal goal of "omic" is the nontargeted analysis of proteins, transcripts and metabolites in a biological sample. The more inspiring phase of omic technology is the sophisticated exploration of assessable dynamics in biological systems. Utilisation of gas chromatography (GC) and liquid chromatography (LC) coupled with mass spectrometry (MS) is enabling high-throughput identification and quantitation of metabolites from various samples. This is a requirement for the analysis of dynamic systems, and hence, metabolomics is one of the key technologies in systems biology area. The data-driven approach is used to discover novel components such as genes, enzymes and metabolites as well as interactions in large-scale omics datasets. Thus, metabolomics aim is to discover new networks based on omics data for functional study [6]. In addition, Saito [7] has demonstrated the use of systems biology approach in photochemistry field for functional genomics study. In this new

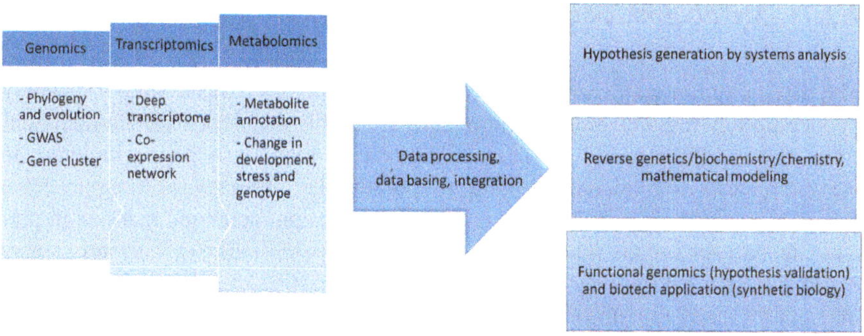

Fig. 4.1 Workflow of data-driven in systems biology and functional genomics. Multi-omics approaches are used to generate hypothesis in systems biology studies. The figure was modified from "Phytochemical genomics—a new trend" of Kazuki Saito [7]

study, the systematic integration of different "omics" was used to investigate regulation and functions of metabolites in plant systems. Hypothesis is generated in this integrated approach and hence will be confirmed by reverse genetics, biochemistry and functional genomics (Fig. 4.1).

4.2 Applications of Metabolomics

Metabolomics can be applied in multifunctional projects utilising diverse types of analytical tools depending on designated research aims [8]. It can be divided into different categories: targeted analysis allowing quantitative analysis of targeted compounds, untargeted metabolic profiling focusing at a qualitative analysis of untargeted metabolites and metabolomics to determine an unbiased overview of metabolic configurations. Collectively, metabolic fingerprinting approach is performed for a rapid screening and determination of metabolites, which could reduce the use of analytical platforms for analysis of metabolites with biochemical relevance [8]. Such technique has become a rapid developing method in postgenomic study [6] or for phenotypic characteristic determination for quick identification and discrimination.

On the other hand, with the recent technological advances in the analytical tools, high-throughput profiling of large numbers of metabolites in biological samples has become common and cheaper. Tools such as mass spectrometry have enabled the simultaneous identification of a broad range and diverse metabolites group. In particular, the application of metabolomics has attracted and currently being used in almost all fields including human disease research and plant and microbial studies. The increment of metabolomics application demonstrates the utility of a metabolomics approach for the understanding of biochemical reactions and its impact.

4.2.1 Metabolomics in Human Studies

The metabolomics application is very important in human disease study especially in oncobiology research areas. Tumour cells are highly proliferated and with higher rates of transcription and translation. Therefore, there is a huge challenge in metabolomics-based medicinal field to predict the presence of tumour cells. Previous studies have focused on cancer detection using metabolic markers in pre-clinical analyses and quantification of the biomarkers in fluids [9]. Currently, many metabolites have been detected and identified, suggesting that they can be used as markers for many tumours (Table 4.1). Subsequently the integrated omics—metabolomics, proteomics and genomics techniques—have accelerated the process of early detection and diagnosis of cancer.

One of the major cases of tumours is the hepatocellular carcinoma (HCC). The identification of cytokine biomarkers using microarrays has been previously

Table 4.1 Metabolites used as biomarkers of human diseases[a]

Disease	Metabolite biomarker	Ref.
Male infertility	Citrate, lactate and glycerylphosphorylcholine	Kovac et al. [16], Hamamah et al. [17]
Lung cancer	Hippurate, trigonelline, β-hydroxyisovalerate, α-hydroxyisobutyrate, N-acetylglutamine and creatinine	Rocha et al. [18]
Alzheimer's disease	Succinic anhydride, pyruvic acid, 2-aminopropanol, n,n-didemethylchlorpromazine, L-alanine, n-butyl ester, L-glutamic acid dibutyl ester, L-dopa, taurine, creatine, creatinine, lactate, β-alanine, cysteine, fumaric acid, 2-octenedioic acid and acetoacetic acid	Xu et al. [19], Trushina et al. [20]
Respiratory diseases	Asthmatic children: acetate Chronic obstructive pulmonary disease (COPD): leucine, lactate, propionate, acetate and pyruvate	Carraro et al. [21], de Laurentiis et al. [22]
Huntington disease	3-Nitropropionic acid	Henry and Mochel [23]
Multiple sclerosis	Elevated levels: 2-aminobutyrate, 1,3-dimethylurrate, glutamate and acetate Reduced levels: oxaloacetate, citrate, alanine and 3-hydroxybutyrate	Smolinska et al. [24]
Impaired glucose tolerance (IGT)	Significantly altered levels: glycine, lysophosphatidylcholine (LPC) (18:2) and acetylcarnitine	Wang-Sattler et al. [25]
Renal cell carcinoma	Phospholipids, phenylalanine, tryptophan, acylcarnitines, cholesterol metabolites and arachidonic acid metabolism	Liu et al. [26]
Colorectal cancer	Acteylcarnine, phenylacetylglutamine, leucylproline and aspartyllysine	Kim et al. [27]
Kidney cancer	Quinolinate, 4-hydroxybenzoate and gentisate	Kim et al. [27]

[a]Adapted from Gomez-Casati et al. [28]

reported [10]. Moreover, some studies have been conducted to profile metabolites in patient with HCC. Patterson et al. utilised UPLC-TOF mass spectrometry and discovered the increased of glycodeoxycholate, deoxycholate-3-sulphate and bilirubin in the patient's biofluid samples [11]. Other polar metabolites such as arginine, alanine and lysine were found to be altered in liver cancer [12]. Other studies also reported that taurine, choline, glycerophosphocholine and phosphocholine were altered in breast cancers. In addition, myo-inositol levels were found to be increased in colon adenocarcinoma, prostate cancer and ovarian carcinoma [9, 13–15].

However, there are several hindrances in the cancer metabolome research. Data analysis difficulty in justifying the group of tumours due to the profile of metabolites could be varied among different types of tumour. The characterisation of metabolites that response to tumour is also difficult due to variation of sample to sample, the accuracy and sensitivity of the analytical systems and the physiological characteristics of the tumour [15].

Many studies have been carried out to search for new metabolic biomarkers for tumour disease. However, the metabolomics approach is still not yet established. In order to understand the metabolic changes and pathway alteration, it is important to integrate omics approaches—metabolomics, proteomics, transcriptomics and genomics. This will provide an appropriate information and better selection of biomarkers for potential application in prevention and diagnosis. The finding will result in a better selection of the potential biomarkers for prevention and diagnosis of the disease [29].

4.2.2 Metabolomics in Plant Studies

Metabolomics is predominantly significant in the plant study, because plants produce a huge number of metabolites—extremely more than are produced by animals and microorganisms [6]. Secondary metabolites that were produced by plant provide many used in response to abiotic stress. Generally, secondary metabolites will be produced by plants in response to the environment stress. Environmental metabolomics is a promising area of study in plant physiology to answer environmental or ecological factors in relation to plant metabolite changes [30, 31]. Metabolomics approaches have been an emerging technology to understanding plant behaviour towards environmental and ecological factors.

The study on plant metabolomics also covers plant responses towards stress and abiotic resistance. Stress in plants involves any significant changes in plant growth conditions that could disturb certain metabolic pathways. Metabolomics could significantly contribute to the stress study in plant biology by identifying metabolites/compounds, stress metabolism by-products, adaptation response of plants and signal transduction of plant molecules. Some examples of plant metabolites associated with biotic and abiotic stresses are polyols mannitol and sorbitol; glycine betaine; sugars such as sucrose, trehalose and fructan; or amino acids. Proline and ectoine involved as osmolytes and osmoprotectant for plant protection under high salt concentration, drought and desiccation stresses [32].

Metabolomics also leads to possible pathway identification of food metabolites for human health benefit. The modification of certain pathway has led to the improvement of plant metabolites with nutritional value. One of the successful stories is the study of genetically modified golden rice (GR), with the β-carotene accumulating in the endosperm [33]. Many metabolites involve in the defence against some human diseases; however, the production level in natural plants is insufficient to bestow optimal benefits. Metabolic engineering was established to enhance the production of important secondary metabolites such as anthocyanins in tomato [34]. The study has reported this new tomato variety has a strong purple coloration with threefold improved antioxidant capability. Moreover, they also fed the cancer-susceptible mice with this tomato variety and proved that this tomato could prolong the lifespan of cancer-susceptible mice.

4.2.3 Microbial Metabolomics

Microbes have impacted our daily life in many ways, giving constant interaction in a delicate and complex interrelationship. Typically microorganism is known for their ability to modulate their metabolic composition to tackle fluctuations in environmental conditions [35]. Changes in the metabolome set of microbes offer not only understanding of phenotypic characteristics but also bioactive compounds that are useful to human and animals. On the other hand, numerous studies have indicated the usefulness of microbes as renewable sources for fine chemical productions [36]. In order to enhance the productivity and yield of the desired compounds, understanding and improvement of the cellular system of particular microbes are required [37]. Microbial metabolomics is one of the many platforms dedicated for the study of microbial system. Microbial metabolomics emphasises on the collection of low-molecular-weight metabolites through separation, identification and quantification using series of analytical platforms from any microbes [38, 39]. Namely, microbial metabolomics tackles two aspects of microbial systems. Firstly, metabolic footprinting or exo-metabolomics that focuses on the excretion of cells into the extracellular surrounding. The second aspect is known as metabolic fingerprinting or intracellular metabolites analysis which deals with the cellular metabolism or metabolites found inside the cell. Both aspects are equally important to provide key information that contributes to the understanding of any microbial systems [40]. However, in comparison with plant metabolomics, the use of metabolomics to study microorganisms is relatively late. Nevertheless, analysis of microbial system by methods of metabolomics has gained increased attention and has considered essential for understanding the cellular functions [41, 42].

E. coli is among the earliest microorganisms that has been used in numerous biotechnology approaches and plays an important role in the development of microbial metabolomics studies [43]. *E. coli* has served as a valuable model for microbial system, and several metabolomics studies aimed at the nontargeted approach to study metabolites in *E. coli* have been reported [44]. Meanwhile lactic acid bacteria

(LAB) have gained significant interests in systems biology studies because of its generally recognized as safe (GRAS) status and extensive usage in fermentation process. More importantly, various beneficial effects to human health such as anti-inflammatory and allergic have been attributed by LAB via its probiotic effects [45]. Specifically, the global metabolomics studies of LAB have been frequently reported [46–48]. For example, the differences in the phenotypic characteristics of *Lactococcus lactis* strains isolated from different environments showed that each strain requires different amino acids for proper growth [49].

Beside *E. coli* and LAB, yeasts such as *Saccharomyces cerevisiae* [50] and filamentous fungi are also known for their impact on human and have been manipulated for the production of bioactive secondary metabolites using methods of metabolomics [51]. Recently, the interest of using filamentous fungi to produce high-valued metabolites has been reported. The various ecological niches and different life styles of filamentous fungi have highlighted the potential of this microorganism to produce a great number of secondary metabolites [52]. For example, the filamentous fungi are source of bulk and fine chemicals of therapeutic compounds that can be used to treat human and plant diseases [53, 54]. However, the global metabolite profiling of yeasts and filamentous fungi has not been frequently described. Furthermore, like other microorganisms, understanding and utilisation of yeasts and fungi require comprehensive strategy such as identification and informatics to increase its usefulness in biotechnology applications.

4.3 Metabolomics Workflow

In the past 10 years, metabolomics has progressively made improvements related to metabolomics software as well as hardware with an increasing complexity of tools and applications. For the discovery study without prior knowledge on the composition of metabolites in the sample, untargeted analytical techniques could be applied to detect a plethora of metabolites. In this case, metabolite identification is crucial following data procurement and processing. Nowadays, the identification of metabolites in untargeted metabolomics studies is a significant bottleneck in metabolomics. To grab the whole concepts of metabolomics, it is important to highlight the workflow of metabolomics.

4.3.1 Experimental Design

Figure 4.2 shows the metabolomics workflow, which starts with a biological question, followed by experimental design, data collection and analysis and finally biological interpretation. Data collection and analysis are the key components in metabolomics workflow. The two components tackle both qualitative and quantitative issues that strengthen the biological interpretation of the final results.

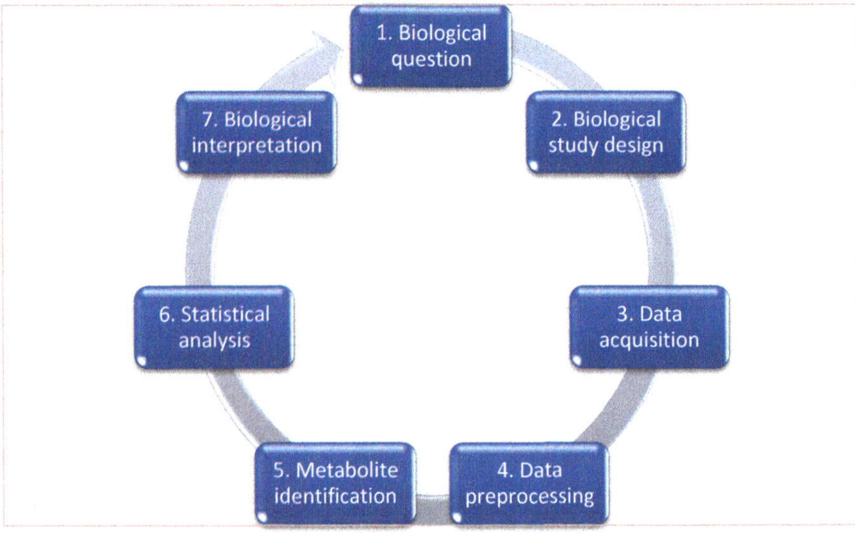

Fig. 4.2 An overview of the workflow of metabolomics

Metabolomics approach is carried out to estimate the effects of treatment and differences between groups and subsequently to know the cause of such differences. Some typical questions that are tackled in metabolomics studies are the following:

- Are there any differences between samples?
- What are the differences between samples?
- What are the reasons for the differences?

To tackle such problems, metabolomics requires maximum information extracted from the subject, taking into account all factors including biasness and variation. Designing an effective experimental workflow to confer both biological questions and variation is one of the vital aspects in metabolomics. Besides that, experimental workflow should also focus on the sample preparation and extraction, which deal with complex biological matrices and sample-size determination.

4.3.2 Data Acquisition

Metabolomics approach has been applied in various fields including medical, synthetic biology, plant, animal and microbial systems. The metabolome represents the changes in phenotype and function in a biological system. Metabolomics studies involve high-resolution analysis using high-throughput analytical tools for simultaneous determination and quantification analysis of metabolites [38]. In particular, the advent of analytical tools has contributed to the generation of biochemical information or metabolome data [55]. The use of analytical tools in metabolomics contributes towards classification and quality control based on the chemical components reported in the chromatogram and mass spectrum [56]. Since the goal of metabolomics is to profile all metabolites in biological samples, metabolomics greatly depends on the application of analytical instruments for acquisition and identification of metabolome data [57]. Gas chromatography-mass spectrometry (GC-MS) is the most favourable analytical tool used in metabolomics studies. GC-MS gives high peak capacity, excellent repeatability, vast and readily available electron ionised (EI) compound libraries and relatively easy instrument handling. GC-MS requires samples to be derivatised or volatilised for detection. Compound identification using GC-MS is straightforward due to the availability of compound library. Identification is based on the generated mass spectrum comparison against standard mass spectrum from other libraries such as the National Institute of Standards and Technology (NIST) mass spectral library.

On the other hand, compared to GC-MS, liquid chromatography-mass spectrometer (LC-MS) detects a wider range of metabolites, from high- to low-molecular-weight compounds as well as hydrophilic and hydrophobic metabolites. LC-MS largely depends on column specification and mobile phase selection for the separation of compounds. Electrospray ionisation (ESI) is generally used for ionisation source in the LC-MS instrument. Moreover, LC-MS does not require derivatisation or special preparation for sample analysis as needed by GC-MS. However, LC-MS gives insufficient peak capacity and stability, mainly due to matrix effect. Despite the disadvantages, LC-MS has been widely opted for simultaneous analysis of both primary and secondary metabolites. Apart from chromatographic techniques, capillary electrophoresis coupled to mass spectrometer (CE-MS) [5] has been introduced to examine metabolites of amino acids, glycolytic system, pentose phosphate pathway and tricarboxylic acid (TCA) cycle. CE-MS is favourable over normal high-pressure liquid chromatography (HPLC) due to its separation capacity. However, CE-MS is also known to have low repeatability and easily influenced by temperature changes. Spectroscopy-based analytical tools such as nuclear magnetic resonance (NMR) and Fourier-transform infrared spectroscopy (FTIR) have also been applied in metabolomics studies. These tools detect specific resonance absorption profiles of metabolites using magnetic field and infrared (IR), respectively [58].

4.3.3 Data Analysis

Data analysis is a crucial process in metabolomics studies. The process aims at finding significant changes and validates the obtained data. Large data or often referred to as metadata are obtained using various analytical platforms and tools. Data analysis strategies in metabolomics studies can be divided into two, namely, the nontargeted and targeted approaches [2]. Additionally, nontargeted may be referred to as chemometric metabolomics where patterns from a sample are processed and significant differences are identified. Targeted approach is mainly referred to as quantitative analysis. Compounds are firstly identified and quantified [59].

4.3.4 Nontargeted Approach

In this step, all possible metabolites are processed prior to identification. Nontargeted or untargeted approach deals with detection of as many groups of metabolites as possible to get patterns or fingerprints. An organised two-dimensional data matrix consisting of metabolites id (retention time (Rt), *m/z* value) and quantitative variables (peak height, peak area, etc.) needs to be generated. To obtain this information, data alignment is required. Alignment is usually carried out using Rt or binning of *m/z* values. In addition, alignment using internal standard Rt may reduce variation among samples and subsequently missing values. Generally, nontargeted approach deals with extremely large data matrices. Therefore, visualisation using multivariate statistical analysis (MVA) is performed to extract interesting and meaningful metabolites from the data matrices. Principal component analysis (PCA) is a non-supervised MVA approach that has been used in data mining of metabolomics. PCA uses score and loading matrices to explain the basis of the variance within the data. By comparing these matrices, the relationship and identification of metabolites that contribute to the differences among samples can be obtained. Apart from that, partial least squares discrimination analysis (PLS-DA) and orthogonal partial least squares discrimination analysis (OPLS-DA) could be used to discriminate important metabolites with biomarker potential [60].

4.3.5 Targeted Approach

Metabolites obtained from nontargeted approach can potentially be false positive or false negative due to analysis procedure and bias. Therefore, extraction of significant metabolites using appropriate methods is required for biological interpretation. On the other hand, targeted approach requires confirmation of significance and reliability of peaks that have been annotated according to its metabolite id. High reproducibility of data matrices using quantification method is important for targeted approach. Quantification can be carried out using calibration curve of pure standard or calculation against internal standard. Similar to nontargeted approach, tools of MVA are also employed for analysis of targeted approach. PCA, PLS-DA and OPLS-DA are used as major MVA tools in metabolomics studies. Finally, in order to obtain the biological interpretation from MVA, additional bioinformatics tools are being integrated into metabolomics studies. Metabolite set enrichment analysis (MSEA), for example, is being utilised to identify and interpret patterns of metabolite changes in biologically meaningful context [61].

4.4 Metabolomics Case Studies

4.4.1 Metabolomics of Kesum Herbs

Saito and Matsuda [6] have described in details how metabolomics technology is emerging and how functional genomics and plant biotechnology can benefit metabolomics technology. Metabolome data driven by systems biology approach have unravelled the secrets of plant cell systems and how it could contribute to the plant biotechnology application. Study on plant species has been carried out in nearly 300,000 species, and yet about 100,000 species still remain unexplored [62]. The overall number of metabolites that exist in plant is about one million [63] showing that it has a plethora of compounds that could be beneficial for pharmaceutical industries. Yet, only a few plants were further explored particularly on their biological activities and chemical constituents of their metabolites [64]. Hence, more research is required in cataloguing the metabolome of unexplored plant species.

Persicaria minor syn. *Polygonum minus* Huds. or locally known as kesum is often used in various traditional medicines, formulations as well as local cuisines. Throughout the years, compounds from the essential oil of this local herb have been successfully profiled. Metabolites produced by this plant were analysed and identified by analytical pipelines, such as gas chromatography-mass spectrometry (GC-MS), GC-MS with flame ionisation detector (GC-FID), two-dimensional gas chromatography time-of-flight mass spectrometry (GC × GCTOF MS) and liquid chromatography time-of-flight mass spectrometry (LC-TOF). A total of 48 compounds using GC × GCTOF MS were successfully identified. By using GC-MS technique, a total of 42 compounds were identified. Hence, the techniques used were also efficient and reliable to identify and quantify the metabolites from *Polygonum minus* Huds [65].

The effects of temperature in regulating major metabolic profiles in the herbal plants have also been reported. A set of simulated reciprocal transplant experiments was performed to understand the temperature effects on plant metabolites. *P. minus* plants were harvested after growing in growth chambers in a controlled environment with the temperature set to mimic lowland and highland conditions. GC-MS and LC-TOF were utilised to identify and classify the metabolites that response to the separation between the treatments. A total of 37 volatile compounds by GC-MS analysis were obtained. Meanwhile, LC-TOF successfully identified a total of 85 flavonoids. Aldehydes and terpene groups were accumulated in highlander's population at treatment of higher temperature. Larger amounts of flavonols were also detected at higher temperature treatment. However, anthocyanin compounds decreased in this treatment. It is possible that the chemical composition was influenced by the effects of temperature on the plant origin [66]. Fourier-transform infrared spectroscopy (FTIR) was also used to characterise different populations of the above-mentioned plant grown in different controlled environments. The plant population was discriminated by using a thermal pertur-

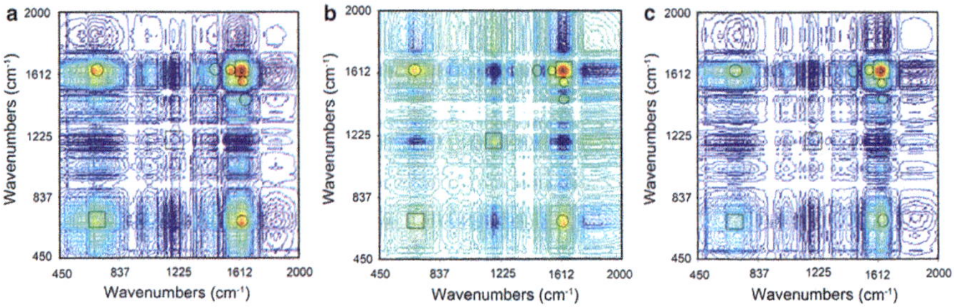

Fig. 4.3 Contour plot of 2D correlation IR synchronous spectroscopy. Two-dimensional spectra from *P. minor* treated in different temperature treatments were analysed in the region of 2,000–450 cm^{-1}. (**a**) Lower growth temperature treatment, (**b**) control, (**c**) higher growth temperature treatment. Square boxes indicate autopeaks. Circles indicate crosspeaks. Red colour indicates strong peak, while light blue indicates weak peak. The figure is reprinted from "Direct discrimination of different plant populations and study on temperature effects by Fourier transform infrared spectroscopy", by K. Khairudin, N. Sukiran, H-H. S.N. Baharum and N. Noor, 2014, *Metabolomics*, p. 203–211. Copyright 2016 by Springer. Adapted with permission

bation technique of 2D-IR correlation spectra (Fig. 4.3). The study implied that IR fingerprinting could directly differentiate the populations of plant origin and also the effect of temperature on the plant's growth [67].

The metabolite profiles were further investigated based on the different tissues of this herbal plant. Techniques such as solid-phase microextraction (SPME) and hydrodistillation were used before the extracts were subjected to GC-MS analysis. The compounds that contribute to the aroma and flavours of this species have been successfully profiled, which were about 77 metabolites. High levels of terpenoids were also detected in leaves, yet much less were detected in stem and root [68].

Even though the volatile compounds from the essential oil of *P. minor* have been intensively studied, information contributed to the aroma-active compounds of this plant species was still lacking. Therefore, the aroma-active compounds of the compelling aroma of *P. minor* were investigated and characterised by using GC-MS/olfactometry (GC-MS/O) and aroma extraction dilution analysis (AEDA) [69]. Based on this finding, several decanal, dodecanal, 1-nonanal, farnesol and α-bergamotene were identified as the key compounds that contributed to the characteristic fragrance of this plant. The finding is important to unravel the information on biosynthesis of aromatic compounds in this herb. Furthermore, this finding also can be further applied in the flavour and fragrance industries.

Further investigation on biological properties of *P. minor* was also carried out [70, 71]. The study was performed by using essential oil and solvent extracts. Antimicrobial, antioxidant and anticholinesterase activities were investigated, and the plant showed high activity towards antioxidants, particularly in DPPH scavenging activity. Meanwhile, aqueous and methanol extracts from leaf samples were shown to have the best acetylcholinesterase inhibitory activity. The highest antimicrobial activity was detected against methicillin-resistant *Staphylococcus aureus* (MRSA). These findings could provide us the first phase of phytochemical profiling for the exploration of the value-added pharma-cognitive properties of this plant species.

4.4.2 *Lactococcus lactis* as a Model for Metabolic Engineering Application

L. lactis has been subjected to several stressful conditions including oxidation, heating and cooling, acid, high osmolality and starvation. Understanding the stress response behaviour is important, not only for strain optimisation but also to manipulate *L. lactis* as host for heterogeneous compound production. Moreover the stress response mechanism in *L. lactis* is of the fundamental interest for metabolic engineering application. *L. lactis* strains are by far the most extensively studied lactic acid bacteria (LAB). Over the last decades, several genetic tools have been developed for this strain. *L. lactis* strains pose an attractive metabolic regulation that suit the development of effective cell factories [72]. *L. lactis* strains are suitable for metabolic engineering manipulation because they do not produce endotoxins and are classified as GRAS. *L. lactis* strains are recognised in two major subspecies: *L. lactis* subsp. *lactis* and *L. lactis* subsp. *cremoris*. Like other LAB, *L. lactis* subsp. *cremoris* is known for its limited biosynthetic capacity to produce essential metabolites and thus explained its complex nutritional requirement [49]. In order to investigate the capability of *L. lactis* subsp. *cremoris* to produce specific metabolites to fulfil its growth requirement, the strain was exposed to different temperatures and agitation. Initially, the extracellular profiling of *L. lactis* cultivated at 30 °C was carried out using Fourier-transform infrared spectroscopy (FTIR). Further investigation was performed on the bacterium that contributes to the production of organoleptic properties by using headspace and gas chromatography-mass spectrometry (HSGC-MS). We found that *L. lactis* subsp. *cremoris* produced high levels of 3-methylbutanal, 2-methylbutanal and 2-methylpropanal. These volatile compounds have been reported to influence the aroma, taste and quality of cheese production. Metabolite profiling was carried out on intra- and extracellular of *L. lactis* when exposed to different conditions. Both temperatures, 37 °C without agitation and 30 °C with agitation (150 rpm), were chosen as the preferred conditions to grow *L. lactis* because these conditions have not been tested for its metabolite profiling but have been explored for transcriptome analysis of *L. lactis* [47].

The growth of *L. lactis* subsp. *cremoris* MG1363 at 30 °C with and without agitation and at 37 °C was monitored using plate counts (cfu/mL) and optical density (OD_{600}). Specifically, the growth curve for all growth conditions can be characterised into mid- exponential, around 3–5 h, early stationary phase after 6–7 h and finally entered stationary phase around 8 h. During exponential phase, the carbon source, glucose, was converted into biomass and fermentation products of lactate, acetate, ethanol and carbon dioxide (CO_2). Decreasing exponential phase took place after 6 h after cultivation. The exponential growth phase is generally considered as the linear part of the growth curve. During this stage, cells are assumed to be in steady state. The exact period of the exponential growth phase determined the right sampling point/time for metabolite and fluxome analysis because, during this stage, all of the intermediate concentration and fluxes are assumed constant. Therefore, in batch cultures, a metabolic steady state exists during exponential growth where the growth rate is constant. In addition, a few studies have shown that exponentially growing cells are much more sensitive to environmental stresses such as during starvation, temperature response, accumulation of acid and ethanol and osmotic and oxidative stresses that their stationary phase counterparts.

Generally, the main factors that affect the characteristics of the metabolic contents in fermented products are the ingredients, fermentation techniques and manufacturing practice. In this study these factors are characterised by the used medium, M17, the tested conditions (30 °C with and without agitation and 37 °C without agitation) and finally detection tools. Nonetheless, the metabolic contents should be strictly associated with the physiochemical composition of the medium. Meanwhile, the environmental adaptability of microbial is resulted from the genetic information, controlled by the complex regulatory networks that enable adaptation to a variety of environments. In this study, 61 metabolites were detected using trimethylsilyl (TMS) derivatisation and 44 metabolites using methyl chloroformate (MCF) derivatisations. However only 47 metabolites were statistically validated ($P < 0.05$), comprising of 13 amino acids, fermentation by-products of acetate and lactate, propanoate metabolism products of acetone and propanoate, butanoate metabolism product of 2,3 butanediol, lipids of butanoic acid, palmitic acid and hexanoic acid.

In order to understand the metabolic changes of *L. lactis* when exposed to different conditions, correlation analysis was performed to investigate the pairwise relationship of profiled metabolites. Generally, metabolites that are biosynthetically linked were clustered together. Figure 4.4 describes the correlation levels between *L. lactis* extracellular with significance ($P < 0.05$). Red colour indicates positive correlation, while green colour exhibits low and negative correlation. Three clusters of metabolites (A, B, C) were obtained from the correlation network. Specifically clusters A and cluster B were linked by butanoic acid. Cluster A was composed of metabolites that are associated with phenylalanine metabolism, such as aniline, benzoate and phenylacetic acid; amino acids and biogenic amines of aspartate, serine, GABA and putrescine; fatty acid group of 1-hexadecanol; and fermentation by-products of lactate and ethanol. Cluster B contained metabolites such as decanoic acid, heptadecane, 2,4-dihydroxyacetophenone and thioglycolic acid. Meanwhile, cluster C comprised of pyruvic acid, threonine and D-arabinono-1,4-lactone together while by-product fermentation of lactate, ethanol and acetate. In particular, metabolites in cluster C were also of interest because the correlated metabolites were identified as by-products of fermentation and may play major contribution towards the phenotypic characteristic of *L. lactis*.

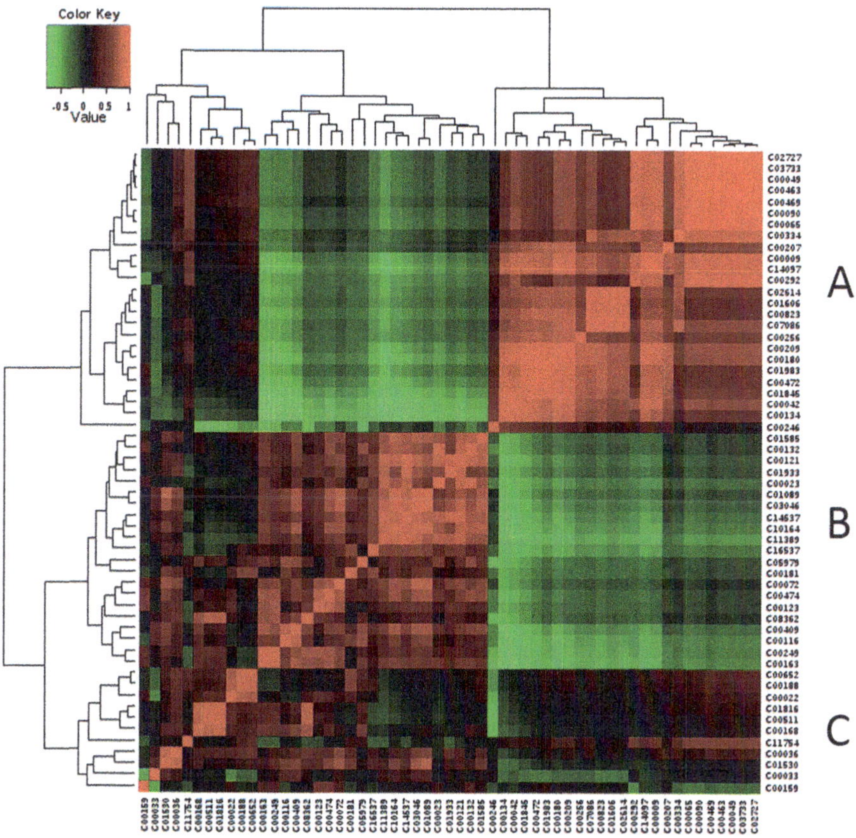

Fig. 4.4 Correlation analysis of the 47 extracellular metabolites ($P < 0.05$) found in all conditions. Green represents correlation coefficient value of >0.7, black represents correlation coefficient value of >0.6 and <0.7 and red represents correlation coefficient value of <0.6. A represents cluster A, B represents cluster B and C represent cluster C

References

1. Oldiges M et al (2007) Metabolomics: current state and evolving methodologies and tools. Appl Microbiol Biotechnol 76:495–511
2. Putri SP et al (2013) Current metabolomics: practical applications. J Biosci Bioeng 115: 579–589
3. Theodoridis G, Gika HG, Wilson ID (2008) LC-MS-based methodology for global metabolite profiling in metabonomics/metabolomics. TrAC Trends Anal Chem 27:251–260
4. Lindon JC, Nicholson JK (2008) Analytical technologies for metabonomics and metabolomics, and multi-omic information recovery. TrAC Trends Anal Chem 27:194–204
5. Weckwerth W, Morgenthal K (2005) Metabolomics: from pattern recognition to biological interpretation. Drug Discov Today 10:1551–1558

6. Saito K, Matsuda F (2010) Metabolomics for functional genomics, systems biology, and biotechnology. Annu Rev Plant Biol 61:463–489 (Merchant S, Briggs WR, Ort D, eds). Annual Reviews. issn 1543–5008, Palo Alto, California, United State of America.
7. Saito K (2013) Phytochemical genomics—a new trend. Curr Opin Plant Biol 16:373–380
8. Fiehn O (2002) Functional genomics (Town C, ed, Ch. 11). Springer, Dordrecht, pp 155–171
9. Serkova NJ, Glunde K (2009) Methods Mol Biol 520:273–295
10. Liu T et al (2011) Rapid determination of serological cytokine biomarkers for hepatitis B virus-related hepatocellular carcinoma using antibody microarrays. Acta Biochim Biophys Sin 43:45–51
11. Patterson AD et al (2011) Aberrant lipid metabolism in hepatocellular carcinoma revealed by plasma metabolomics and lipid profiling. Cancer Res 71:6590–6600
12. Chen J et al (2009) Metabonomics study of liver cancer based on ultra performance liquid chromatography coupled to mass spectrometry with HILIC and RPLC separations. Anal Chim Acta 650:3–9
13. Lee JH et al (2011) P117. H+-myo-inositol transporter SLC2A13 as a potential marker for cancer stem cells in an oral squamous cell carcinoma. Oral Oncol 47(Supplement 1):S111
14. Serkova NJ et al (2008) The metabolites citrate, myo-inositol, and spermine are potential age-independent markers of prostate cancer in human expressed prostatic secretions. Prostate 68:620–628
15. Spratlin JL, Serkova NJ, Eckhardt SG (2009) Clinical applications of metabolomics in oncology: a review. Clin Cancer Res 15:431–440
16. Kovac JR, Pastuszak AW, Lamb DJ (2013) The use of genomics, proteomics, and metabolomics in identifying biomarkers of male infertility. Fertil Steril 99:998–1007
17. Hamamah S et al 1H nuclear magnetic resonance studies of seminal plasma from fertile and infertile men. Int J Gynecol Obstet 43:96–97
18. Rocha CM et al (2011) Metabolic signatures of lung cancer in biofluids: NMR-based metabonomics of blood plasma. J Proteome Res 10:4314–4324
19. Xu X-H, Huang Y, Wang G, Chen S-D (2012) Metabolomics: a novel approach to identify potential diagnostic biomarkers and pathogenesis in Alzheimer's disease. Neurosci Bull 28:641–648
20. Trushina E, Dutta T, Persson X-MT, Mielke MM, Petersen RC (2013) Identification of altered metabolic pathways in plasma and CSF in mild cognitive impairment and Alzheimer's disease using metabolomics. PLOS ONE 8:e63644
21. Carraro S et al (2007) Metabolomics applied to exhaled breath condensate in childhood asthma. Am J Respir Crit Care Med 175:986–990
22. de Laurentiis G et al (2008) Metabonomic analysis of exhaled breath condensate in adults by nuclear magnetic resonance spectroscopy. Eur Respir J 32:1175–1183
23. Henry PG, Mochel F (2012) The search for sensitive biomarkers in presymptomatic Huntington disease. J Cereb Blood Flow Metab 32:769–770
24. Smolinska A, Blanchet L, Buydens LMC, Wijmenga SS (2012) NMR and pattern recognition methods in metabolomics: From data acquisition to biomarker discovery: A review. Anal Chim Acta 750:82–97
25. Wang-Sattler R et al (2012) Novel biomarkers for pre-diabetes identified by metabolomics. Mol Syst Biol 8:1–11
26. Liu G, Snapp HM, Ji QC, Arnold ME (2009) Strategy of accelerated method development for high-throughput bioanalytical assays using ultra high-performance liquid chromatography coupled with mass spectrometry. Anal Chem 81:9225–9232
27. Kim K et al (2011) Urine metabolomic analysis identifies potential biomarkers and pathogenic pathways in kidney cancer. OMICS 15:293–303
28. Gomez-Casati DF, Zanor MI, Busi MV (2013) Metabolomics in plants and humans: applications in the prevention and diagnosis of diseases. Biomed Res Int 2013:792527
29. Herder C, Karakas M, Koenig W (2011) Biomarkers for the prediction of type 2 diabetes and cardiovascular disease. Clin Pharmacol Ther 90:52–66

30. Brunetti C, George RM, Tattini M, Field K, Davey MP (2013) Metabolomics in plant environmental physiology. J Exp Bot 64:4011–4020
31. Viant MR, Sommer U (2012) Mass spectrometry based environmental metabolomics: a primer and review. Metabolomics 9:144–158
32. Mittler R (2002) Oxidative stress, antioxidants and stress tolerance. Trends Plant Sci 7:405–410
33. Paine JA et al (2005) Improving the nutritional value of Golden Rice through increased pro-vitamin A content. Nat Biotech 23:482–487
34. Butelli E et al (2008) Enrichment of tomato fruit with health-promoting anthocyanins by expression of select transcription factors. Nat Biotech 26:1301–1308
35. Merlo ME, Jankevics A, Takano E, Breitling R (2011) Exploring the metabolic state of micro-organisms using metabolomics. Bioanalysis 3:2443–2458
36. Gaspar P, Carvalho AL, Vinga S, Santos H, Neves AR (2013) From physiology to systems metabolic engineering for the production of biochemicals by lactic acid bacteria. Biotechnol Adv 31:764–788
37. Toya Y, Shimizu H (2013) Flux analysis and metabolomics for systematic metabolic engineering of microorganisms. Biotechnol Adv 31:818–826
38. Mozzi F, Ortiz ME, Bleckwedel J, De Vuyst L, Pescuma M (2013) Metabolomics as a tool for the comprehensive understanding of fermented and functional foods with lactic acid bacteria. Food Res Int 54:1152–1161
39. Zhang W, Li F, Nie L (2010) Integrating multiple 'omics' analysis for microbial biology: application and methodologies. Microbiology 156:287–301
40. Mapelli V, Olsson L, Nielsen J (2008) Metabolic footprinting in microbiology: methods and applications in functional genomics and biotechnology. Trends Biotechnol 26:490–497
41. Liebeke M, Dörries K, Meyer H, Lalk M (2012) Functional genomics: methods and protocols (Kaufmann M, Klinger C, eds). Springer, New York, pp 377–398
42. Mashego MR et al (2006) Microbial metabolomics: past, present and future methodologies. Biotechnol Lett 29:1–16
43. Rabinowitz JD (2007) Cellular metabolomics of *Escherichia coli*. Expert Rev Proteomics 4:187–198
44. Winder CL et al (2008) Global metabolic profiling of *Escherichia coli* cultures: an evaluation of methods for quenching and extraction of intracellular metabolites. Anal Chem 80:2939–2948
45. Ménard S et al (2004) Lactic acid bacteria secrete metabolites retaining anti-inflammatory properties after intestinal transport. Gut 53:821–828
46. Azizan KA, Baharum SN, Mohd Noor N (2012) Metabolic profiling of *Lactococcus lactis* under different culture conditions. Molecules 17:8022
47. Taibi A, Dabour N, Lamoureux M, Roy D, LaPointe G (2011) Comparative transcriptome analysis of *Lactococcus lactis* subsp. cremoris strains under conditions simulating Cheddar cheese manufacture. Int J Food Microbiol 146:263–275
48. Tan-a-ram P et al (2011) Assessment of the diversity of dairy *Lactococcus lactis* subsp. *lactis* isolates by an integrated approach combining phenotypic, genomic, and transcriptomic analyses. Appl Environ Microbiol 77:739–748
49. Ayad EHE, Verheul A, de Jong C, Wouters JTM, Smit G (1999) Flavour forming abilities and amino acid requirements of *Lactococcus lactis* strains isolated from artisanal and non-dairy origin. Int Dairy J 9:725–735
50. Schneider K et al (2009) Metabolite profiling studies in *Saccharomyces cerevisiae*: an assisting tool to prioritize host targets for antiviral drug screening. Microb Cell Factories 8:1–14
51. Smedsgaard J, Nielsen J (2005) Metabolite profiling of fungi and yeast: from phenotype to metabolome by MS and informatics. J Exp Bot 56:273–286
52. Kluger B, Lehner S, Schuhmacher R (2015) Metabolomics and secondary metabolite profiling of filamentous fungi. In: Zeilinger S, Martín J-F, García-Estrada C (eds) Biosynthesis and molecular genetics of fungal secondary metabolites, vol 2. Springer, New York, pp 81–101
53. Barkal LJ et al (2016) Microbial metabolomics in open microscale platforms. Nat Commun 7:1–11

54. Thrane U, Anderson B, Frisvad JC, Smedsgaard J (2007) Metabolomics: a powerful tool in systems biology (Nielsen J, Jewett MC, eds). Springer, Berlin/Heidelberg, pp 235–252
55. Zurbriggen MD, Moor A, Weber W (2012) Plant and bacterial systems biology as platform for plant synthetic bio(techno)logy. J Biotechnol 160:80–90
56. Zhao Y et al (2011) Tentative identification, quantitation, and principal component analysis of green pu-erh, green, and white teas using UPLC/DAD/MS. Food Chem 126:1269–1277
57. Putri SP, Yamamoto S, Tsugawa H, Fukusaki E (2013) Current metabolomics: technological advances. J Biosci Bioeng 116:9–16
58. Mouwen DJM, Hörman A, Korkeala H, Alvarez-Ordóñez A, Prieto M (2011) Applying Fourier-transform infrared spectroscopy and chemometrics to the characterization and identification of lactic acid bacteria. Vib Spectrosc 56:193–201
59. Xia J, Wishart DS (2010) MSEA: a web-based tool to identify biologically meaningful patterns in quantitative metabolomic data. Nucleic Acids Res 38:W71–W77
60. Zeng M et al (2010) Plasma metabolic fingerprinting of childhood obesity by GC/MS in conjunction with multivariate statistical analysis. J Pharm Biomed Anal 52:265–272
61. Xia J, Wishart DS (2010) MSEA: web-based tool to identify biologically meaningful patterns in quantitative metabolomic data. Nucleic Acids Res 38:71–77
62. Sticher O (2008) Natural product isolation. Nat Prod Rep 25:517–554
63. Dixon RA, Strack D (2003) Phytochemistry meets genome analysis, and beyond. Phytochemistry 62:815–816
64. Yamada T et al (2008) Mutation of a rice gene encoding a phenylalanine biosynthetic enzyme results in accumulation of phenylalanine and tryptophan. Plant Cell 20:1316–1329
65. Baharum SN, Bunawan H, Ghani MA, Wan Aida Wan M, Noor NM (2010) Analysis of the chemical composition of the essential oil of *Polygonum minus* Huds. Using two-dimensional gas chromatography-time-of-flight mass spectrometry (GC-TOF MS). Molecules 15:7006–7015
66. Goh HH, Khairudin K, Sukiran NA, Normah MN, Baharum SN (2016) Metabolite profiling reveals temperature effects on the VOCs and flavonoids of different plant populations. Plant Biol 18:130–139
67. Khairudin K, Sukiran N, Goh H-H, Baharum S, Noor N (2013) Direct discrimination of different plant populations and study on temperature effects by Fourier transform infrared spectroscopy. Metabolomics 10:203–211
68. Ahmad R et al (2014) Volatile profiling of aromatic traditional medicinal plant, *Polygonum minus* in different tissues and its biological activities. Molecules 19:19220
69. Azizun Rusdi N, Goh HH, Baharum S (2016) GC-MS/Olfactometric characterisation and aroma extraction dilution analysis of aroma active compounds in *Polygonum minus* essential oil. Plant Omics 9:289
70. Hassim N et al (2015) Antioxidant and antibacterial assays on *Polygonum minus* extracts: different extraction methods. Int J Chem Eng 2015:10
71. Hassim N, Markom M, Anuar N, Baharum SN (2014) Solvent selection in extraction of essential oil and bioactive compounds from *Polygonum minus*. J Appl Sci 14:1440–1444
72. Neves AR, Pool WA, Kok J, Kuipers OP, Santos H (2005) Overview on sugar metabolism and its control in *Lactococcus lactis*—the input from in vivo NMR. FEMS Microbiol Rev 29:531–554

Chapter 5
Integrative Multi-Omics Through Bioinformatics

Hoe-Han Goh

Abstract This chapter introduces different aspects of bioinformatics with a brief discussion in the systems biology context. Example applications in network pharmacology of traditional Chinese medicine, systems metabolic engineering, and plant genome-scale modelling are described. Lastly, this chapter concludes on how bioinformatics helps to integrate omics data derived from various studies described in previous chapters for a holistic understanding of secondary metabolite production in *P. minus*.

Keywords Genome-scale model · Omics integration · Multi-omics · Network analysis · Systems metabolic engineering

5.1 Introduction

The overwhelming trend in omics studies relies heavily on bioinformatics to store, mine, process, analyse, interpret, and curate biological big data. Bioinformatics includes computer science, statistics, and mathematical methods, with computer programming for the analysis of various sequence data in molecular biology. The term bioinformatics was introduced in 1970 for the study of biosystems information processes, which has evolved into an interdisciplinary field largely dealing with computational methods for comparative genomic data analysis since the late 1980s [1]. In general, bioinformatics refers to biological studies aided by computer programming apart from data analysis pipelines, especially in the field of genomics such as that of illustrated in previous chapters.

H.-H. Goh (✉)
Institute of Systems Biology, Universiti Kebangsaan Malaysia (UKM), Bangi, Malaysia
e-mail: gohhh@ukm.edu.my

W. M. Aizat et al. (eds.), *Omics Applications for Systems Biology*,
Advances in Experimental Medicine and Biology 1102,
https://doi.org/10.1007/978-3-319-98758-3_5

5.2 Different Aspects of Bioinformatics

Bioinformatics covers many aspects of fundamental and applied research, from hypothesis-driven to data-driven (Fig. 5.1). The hypothesis-driven bottom-up approach is largely knowledge based and depends strongly on modelling and computational simulation for understanding of biological processes. For example, mathematical modelling of enzyme kinetics in a reaction pathway or simulation of flux distribution in a genome-scale model can help identify rate-limiting enzyme/metabolite [2, 3].

On the other hand, data-driven bioinformatics evolved in the mid-1990s as demanded by the Human Genome Project, which led to the explosion of high-throughput omics data. The advancement in sequencing technology dominates the development of bioinformatics, for the acquisition, analysis, and management of tremendous volume of biological data. This is paralleled by the advancement of information technologies, algorithms, and computational and statistical methods. Computationally intensive techniques, such as data mining [4], machine learning, visualisation [5], and pattern recognition, are indispensable with continuous improvement of bioinformatics software and tools for efficient access, analysis, and curation of heterogeneous datasets. Bioinformatics even encompasses solving problems arising from database management. Common sequence analyses include sequence alignment, genome assembly, gene prediction, and functional annotation, as compared to gene and protein expression studies which are based on abundance analysis, in which the latter relies on mass spectrometry for protein fragment identification. Image analysis involves important automated techniques for the microscopic tracing of subcellular molecular movement, as well as phenotypic tracking of

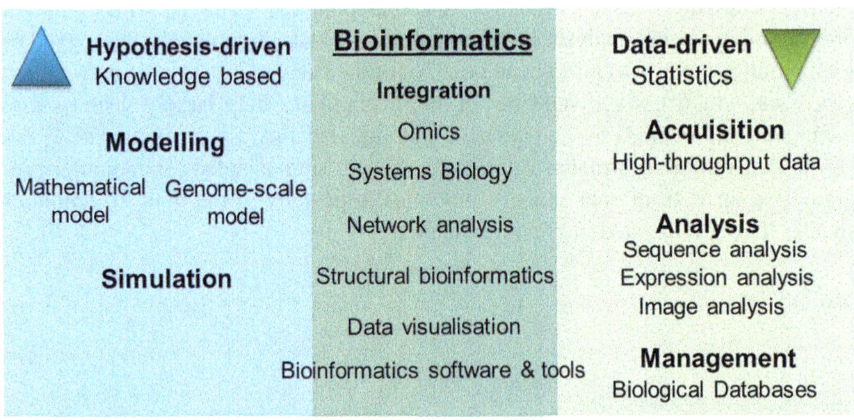

Fig. 5.1 An overview of different aspects of bioinformatics, from knowledge-based hypothesis-driven bottom-up approach to top-down statistics and data-driven

organ growth in real time. Protein structure prediction is a field of structural bioinformatics important for the inference of structure-function relationship to understand the molecular mechanism or protein-protein/metabolite interactions, which can be applied for drug design.

Nowadays, the field of bioinformatics is largely data-driven. Computational modelling and simulation in network analysis have become increasingly important for the integration of multi-omics in the context of systems biology. Table 5.1 summarises the different aspects of bioinformatics.

Table 5.1 Different aspects of bioinformatics with examples

Aspect	Example
Sequence analysis	DNA/RNA sequencing Sequence assembly Genome annotation Genetics of disease Cancer genomics (oncogenomics) Comparative genomics Genetics and population analysis Computational evolutionary biology
Expression analysis	Gene expression Protein expression Metabolite profiling Analysis of expression regulation
Structural bioinformatics	Genome modelling (3D chromatin) RNA secondary structure prediction Protein structure prediction Homology modelling Structure-based drug/chemical design Molecular docking
Image analysis	Microscopy image analysis Automated cell tracking Pattern recognition Bioimage annotation Visualisation
Data management	Database development Web services Curation Meta-analysis Workflow management system
Network and systems biology	Biological network analysis Gene co-expression analysis Genome-scale modelling Molecular interaction networks
Others	Ontology and data integration Literature/text mining Software/tool development

5.3 Bioinformatics for Systems Biology

Essentially, systems biology constitutes a crossover between knowledge-based modelling and omics data-driven approaches. Bioinformatics is a broad multidisciplinary field which is indispensable for systems biology that deals with omics data, mathematical modelling, and network analysis. This is because the dynamic behaviours of biological systems are beyond human intuitive grasp due to the sheer number of components (biomolecules, cells, drugs, and each other) which interact. System-level understanding is only possible through computational models and simulations. Metabolic, gene regulatory, and protein-protein interaction networks are the core of common systems studies, with many examples in *E. coli* [6, 7] and yeast [8–10]. Detailed descriptions and discussion are beyond the scope of this chapter. Readers can refer to recent literature [11–13] to understand further the bioinformatics tools available for systems biology.

5.4 Applications of Bioinformatics

In this section, examples of bioinformatics applications on integrative omics are described for molecular medicine, systems metabolic engineering, and plant genome-scale modelling.

5.4.1 *Integrative Omics in Network Pharmacology*

Network pharmacology is a new paradigm in postgenomic era of molecular medicine for drug design or discovery [14]. This is based on the realm that one drug often targets many proteins and one protein can be targeted by many drugs. Hence, a combination of different drugs could be beneficial synergistically in treating complex diseases. This also led to the current trend of drug repositioning/repurposing, whereby known drugs/compounds are applied for treatment of new diseases.

Network pharmacology relies on a multi-omics systems biology approach, which analyses various omics data together using bioinformatics tools [5, 15] to develop disease networks, drug-target networks, or drug-disease networks [16, 17]. One good example is the use of this approach to discover multicomponent drugs from traditional Chinese medicine (TCM) for multi-target therapy [18–20]. To achieve this, TCM pattern in a disease can be identified using molecular network biomarkers and integrate with pharmacological network of herbal formulas (Fig. 5.2).

The construction of disease-TCM pattern molecular network depends on multi-omics data analysis of categorised patients, according to TCM pattern based on expert consensus or literature analysis. Text mining of SinoMed database helps identify TCM herbal combinations for the treatment of disease with specific TCM

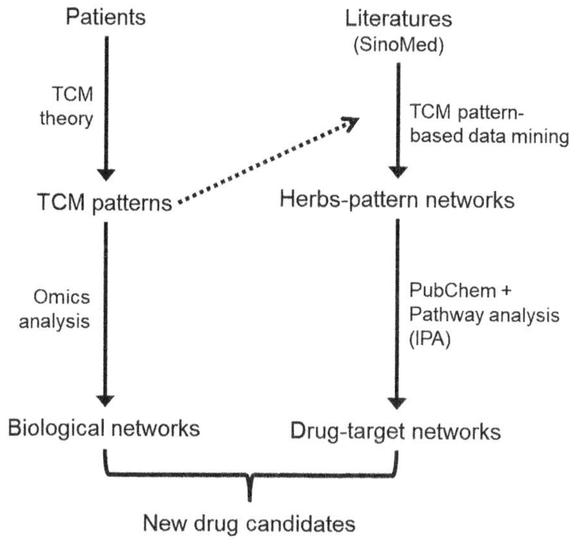

Fig. 5.2 A conceptual framework of network pharmacology for multiple compound drug discovery from TCM

patterns. Targeted proteins by the active compounds in the TCM herbal formula obtained from PubChem are used to construct drug-target networks. Potential multiple-compound drug candidates can then be shortlisted from well-matched compound combinations between disease-TCM pattern molecular network and pharmacological network of herbal formulas. This is not possible through reductionist approach in the past without systems approach of network analysis which requires computing resources. A good example of TCM drug repositioning is reported recently on the use of systems pharmacology approach in the discovery of Liuweiwuling therapeutic use for liver failure [21].

5.4.2 Integrative Omics for Systems Metabolic Engineering

The emergence of ethnomedicine as alternatives of disease treatment has increase the demands for natural products and bioactive compounds as drugs [22], For example, an antimalarial drug artemisinin from a TCM *Artemisia annua* has driven engineered production of its precursor artemisinic acid in yeast [23].

There is a growing trend of employing synthetic biology approach for genetically engineering metabolic pathways in microbial system to produce natural and synthetic compounds. For this purpose, bioinformatics plays a key role in the selection, synthesis, assembly, and optimisation of the parts (enzymes and regulatory elements), devices (pathways), and systems [24]. Furthermore, systems metabolic engineering often employs genome-scale models for flux analysis of the metabolic reconstruction [25]. Hence, fluxomics play important role for optimising flux distribution towards target compound production. Genome-scale metabolic reconstructions allow the modelling on the effects of gene knockouts. However, this is largely dominated by microbes such as *E. coli* and *S. cerevisiae*. Much of the curated/predicted metabolic reconstructions can be found at MetaCyc and BioCyc databases [26]; see http://systemsbiology.ucsd.edu/InSilicoOrganisms/OtherOrganisms for an updated list. This systems approach has accelerated the development of metabolic engineering, such as that of the use of *E. coli* for the production of terpenoids [27] and bioethanol [28].

Recently, multi-omics has become a common approach for comprehensive understanding of different microbial strains by compensating each omics' limitations as illustrated in Fig. 5.3. The ultimate aim is to improve titre, yield, and productivity of engineered microbial cell factories. For that purpose, multi-omics systems biology contributes in the understanding of cellular metabolic status, genome-wide identification of knockout or overexpression targets, pathway prediction, and even enzyme design through computational structural prediction. Further descriptions and discussion on systems metabolic engineering with the integration of systems and synthetic biology with evolutionary engineering can refer to the next chapter and a recent review [29] with references therein. Fondi and Liò (2015) provide a good review for tools used in integrating multi-omics for metabolic modelling pipelines [30].

5.4.3 Integrative Omics for Genome-Scale Modelling in Plants

As mentioned above, genome-scale metabolic model (GEM) is an in silico metabolic flux model constructed from genome annotation-derived metabolic networks with stoichiometry of all known metabolic reactions. GEM is often built by algorithms with constraint-based flux (reaction rate) analysis within defined system boundaries to bridge between modelled metabolic network structure and observed metabolic processes. Constraints are important to limit possible flux values (solution space) in the studied system, which include mass balancing, physico-thermochemical, and actual flux measurements [31]. Flux balance analysis (FBA) is the most popular mathematical method for the phenotypic solution space exploration through linear programming.

GEM allows the assessment of the essentiality of metabolic steps. This enables the prediction of gene targets for knockout or overexpression and is useful for flux optimisation and designing rational metabolic engineering strategies, especially for microbial systems. It is more challenging to construct GEM for higher organisms, especially plants due to complexity of plant cells with photosynthesis/photorespiration, compartmentation, tissue differentiation, diverse metabolic processes, and responses to endogenous (phytohormones) and environmental stimuli [31]. The first ever plant GEM was reported in 2009 for *Arabidopsis thaliana* cell suspension cultures [32]. Other selected examples and their significance are provided in Table 5.2. Previously neglected secondary metabolism is also gaining momentum with the latest advancement of omics approaches in filling in the gaps of metabolomics and proteomics data, especially in medicinal plants producing important bioactive compounds [33].

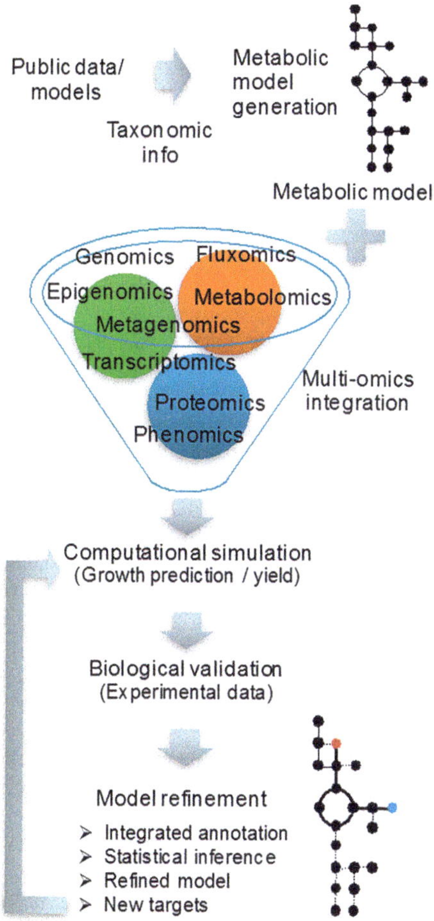

Fig. 5.3 Overall framework of a metabolic modelling/reconstruction pipeline with multi-omics integration, computational simulation, biological validation, and iterative model refinement. Pre-existing genome annotation in public repositories provides information on the presence/absence of metabolic pathways and overall metabolic capabilities of a microbe. For a novel microbe, a metabolic model can be generated from the closest related species with publicly available data based on taxonomic information or from de novo genome sequencing and assembly. Next, different layers of datasets resulting from the application of different omics technologies can be integrated for computational simulation of phenotype prediction. Multi-omics integration provides a more comprehensive perspective on the microbe under study, statistically grounded inferences, novel questions to be addressed, or new target genes to be manipulated, possibly through reiterating the pipeline based on experimental data for further refinement of model.

Despite that GEM is now possible in plants, challenges remain on filling in missing metabolic information with the integration of regulatory and signalling components in dynamic simulation. In this respect, multi-condition, single-platform omics studies such as transcriptomics will be useful for mapping gene expression

Table 5.2 Selected examples of plant GEMs

Species	GEM	Significance	Reference
Arabidopsis thaliana	Suspension cell culture	The first GEM of a heterotrophic plant cell derived from AraCyc with 855 mass-balanced reactions	Poolman et al. [32]
	AraGEM (*i*RS1597)	Compartmented C_3 mesophyll cell model of photosynthesis, photorespiration, and heterotrophic metabolism	Dal'Molin et al. [34]
Brassica napus	Developing embryo	Compartmented FBA model to predict the regulation of oil biosynthesis	Pilalis et al. [35]
	bna572 (oilseed)	Flux balance and variability analyses based on highly comprehensive and compartmentalised oilseed model with in silico mutant analysis in relation to carbon use efficiency	Hay and Schwender [36, 37]
Hordeum vulgare	Seed endosperm	Tissue-specific compartmented FBA model with 65 transporters for study of hypoxia and aerobic conditions	Grafahrend-Belau et al. [38], Rolletschek et al. [39]
Zea mays, *Saccharum officinarum*, *Sorghum bicolor*	C4GEM	Compartmented two-cell (mesophyll and bundle sheath cells) model representing C_4 plants with 112 transporters	Dal'Molin et al. [40]
Zea mays	*i*RS1563	Expanded C4GEM model with secondary metabolism validated using lignin biosynthetic mutants	Saha et al. [41]
Oryza sativa	General	First model in rice with 1736 reactions to study metabolism under varying light intensity	Poolman et al. [42]
	*i*OS2164	A fully compartmentalised model with multi-omics analysis	Lakshmanan et al. [43]
Solanum lycopersicum	*i*HY3410 leaf	Compartmentalised metabolic model of leaf with five organelles to describe metabolic response to drought, particularly photorespiratory metabolism	Yuan et al. [44]
Vitis vinifera	Suspension cell culture	Established cell energy, redox status, and α-ketoglutarate availability as metabolic drivers for anthocyanin accumulation under nitrogen limitation	Soubeyrand et al. [45]

data onto GEM to generate condition-specific models for more realistic depiction of actual metabolic states. Similarly, quantitative proteomics can also be applied for modelling system-level metabolic changes following experimental perturbations, assuming that gene expression or protein abundance correlates with metabolic fluxes. Incorporating multi-conditions transcriptomics and proteomics data will enable condition-based simulation with the elements of gene/protein regulation in switching a pathway on/off. Lastly, metabolomics profiling under different conditions allows the comprehensive identification of metabolite compositional changes to narrow down target pathways for further fluxomics analysis (^{13}C-based) under different experimental conditions. With multi-omics, multi-conditions data, a more realistic dynamic GEM can be simulated to predict outcomes for various scenarios. In plants, GEMs of different tissues, such as root to shoot, can be integrated for whole-plant simulation [46]. With the integration of regulation into GEMs, we can gain important insights of plant metabolic plasticity for rational metabolic engineering to improve plant biomass production through higher tolerance and resistance to biotic and abiotic stresses.

5.5 Case Study: Integrating Multi-Omics in *Polygonum minus*

Over the past 10 years, extensive studies using different omics approaches have been performed on aromatic herb *Polygonum minus* as described in previous chapters. Much is learnt about *P. minus* on the transcriptomes [47–49] and metabolomes [50–52] from different tissues, as well as molecular responses towards elicitors [53–55]. The integration between transcriptomics and metabolomics studies [56] allows the reconstruction of secondary metabolite biosynthetic pathways. This also helps in the elucidation of global gene reprogramming which resulted in the compositional changes of volatile organic compounds (VOCs) in response to elicitation or other environmental factors. Furthermore, the established transcriptome sequences provide a reference for the identification of proteins in shotgun proteomics through proteomics informed by transcriptomics (PIT) approach [57].

General research framework of integrating multi-omics results in *P. minus* is shown in Fig. 5.4. This is applicable for other plants/organisms without a reference genome, particularly tropical medicinal plants, which have scarce sequence information and limited knowledge on the production of bioactive compounds. By eluci-

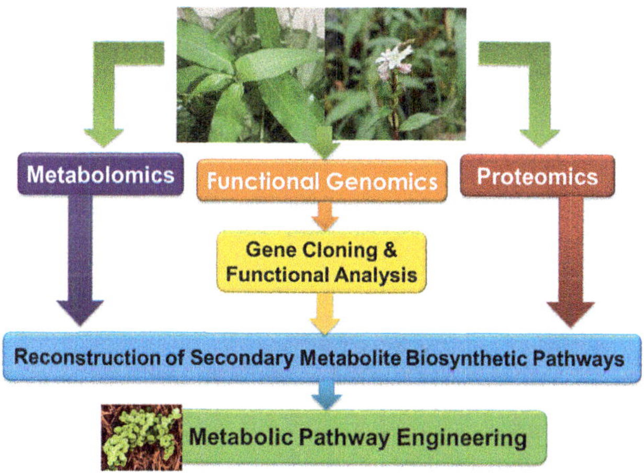

Fig. 5.4 Research framework for the integration of multi-omics studies in *P. minus*

dating the genes and enzymes involved in pathways of secondary metabolite biosynthesis, metabolic engineering in microbial system becomes possible through synthetic biology approach (described in the next chapter). Hence, integrative omics through systems biology approach provides a fundamental blueprint to enable applied large-scale production of targeted compounds through microbial bioengineering.

References

1. Hogeweg P (2011) The roots of bioinformatics in theoretical biology. PLOS Comput Biol 7:e1002021
2. Henry CS et al (2010) High-throughput generation, optimization and analysis of genome-scale metabolic models. Nat Biotechnol 28:977–982
3. Oberhardt MA, Palsson BØ, Papin JA (2009) Applications of genome-scale metabolic reconstructions. Mol Syst Biol 5:320
4. Prasad TV, Ahson SI (2007) Bioinformatics: applications in life and environmental sciences. Springer Netherlands Capital Publishing Company, New Delhi, India. pp 145–172
5. Tao Y, Liu Y, Friedman C, Lussier YA (2004) Information visualization techniques in bioinformatics during the postgenomic era. Drug Discov Today BIOSILICO 2:237–245
6. Shen-Orr SS, Milo R, Mangan S, Alon U (2002) Network motifs in the transcriptional regulation network of *Escherichia coli*. Nat Genet 31:64–68
7. Feist AM et al (2007) A genome-scale metabolic reconstruction for *Escherichia coli* K-12 MG1655 that accounts for 1260 ORFs and thermodynamic information. Mol Syst Biol 3:121
8. Duarte NC, Herrgård MJ, Palsson BØ (2004) Reconstruction and validation of *Saccharomyces cerevisiae* iND750, a fully compartmentalized genome-scale metabolic model. Genome Res 14:1298–1309
9. Lee TI et al (2002) Transcriptional regulatory networks in *Saccharomyces cerevisiae*. Science 298:799–804
10. Schwikowski B, Uetz P, Fields S (2000) A network of protein-protein interactions in yeast. Nat Biotechnol 18:1257–1261
11. Krawetz S (2009) Bioinformatics for systems biology. Humana Press, Totowa
12. Likić VA, McConville MJ, Lithgow T, Bacic A (2010) Systems biology: the next frontier for bioinformatics. Adv Bioinforma 2010:1
13. Tran QN, Arabnia HR (2016) Emerging trends in applications and infrastructures for computational biology, bioinformatics, and systems biology: systems and applications. Elsevier/Morgan Kaufmann, Amsterdam/Boston
14. Tang J, Aittokallio T (2014) Network pharmacology strategies toward multi-target anticancer therapies: from computational models to experimental design principles. Curr Pharm Des 20:23–36
15. Valencia A (2002) Bioinformatics and computational biology at the crossroads of post-genomic technology. Phytochem Rev 1:209–214
16. Ostrowski J (2008) Integrative genomics – a basic and essential tool for the development of molecular medicine. Acta Pol Pharm Drug Res 65:621–624
17. Yan Q (2013) Handbook of personalized medicine: advances in nanotechnology, drug delivery and therapy. Pan Stanford, New York, pp 191–220
18. Hao DC, Xiao PG (2014) Network pharmacology: A Rosetta stone for traditional Chinese medicine. Drug Dev Res 75:299–312
19. Li S, Zhang B (2013) Traditional Chinese medicine network pharmacology: theory, methodology and application. Chin J Nat Med 11:110–120

20. Tao WY, Wang LY, Huang GQ, Luo M (2013) *Applied mechanics and materials*, vol 411–414. Trans Tech Publications Ltd., Durnten-Zurich, pp 3141–3145
21. Wang J-B et al (2018) A systems pharmacology-oriented discovery of a new therapeutic use of the TCM formula Liuweiwuling for liver failure. Sci Rep 8:5645
22. Li JWH, Vederas JC (2009) Drug discovery and natural products: end of an era or an endless frontier? Science 325:161–165
23. Ro DK et al (2006) Production of the antimalarial drug precursor artemisinic acid in engineered yeast. Nature 440:940–943
24. Carbonell P et al (2016) Bioinformatics for the synthetic biology of natural products: integrating across the Design-Build-Test cycle. Nat Prod Rep 33:925–932
25. Blazeck J, Alper H (2010) Systems metabolic engineering: genome-scale models and beyond. Biotechnol J 5:647–659
26. Caspi R et al (2009) The MetaCyc database of metabolic pathways and enzymes and the BioCyc collection of pathway/genome databases. Nucleic Acids Res 38:D473–D479
27. Martin VJJ, Piteral DJ, Withers ST, Newman JD, Keasling JD (2003) Engineering a mevalonate pathway in *Escherichia coli* for production of terpenoids. Nat Biotechnol 21:796–802
28. Yim H et al (2011) Metabolic engineering of *Escherichia coli* for direct production of 1,4-butanediol. Nat Chem Biol 7:445–452
29. Chae TU, Choi SY, Kim JW, Ko Y-S, Lee SY (2017) Recent advances in systems metabolic engineering tools and strategies. Curr Opin Biotechnol 47:67–82
30. Fondi M, Liò P (2015) Multi -omics and metabolic modelling pipelines: challenges and tools for systems microbiology. Microbiol Res 171:52–64
31. Collakova E, Yen JY, Senger RS (2012) Are we ready for genome-scale modeling in plants? Plant Sci 191–192:53–70
32. Poolman MG, Miguet L, Sweetlove LJ, Fell DA (2009) A genome-scale metabolic model of Arabidopsis and some of its properties. Plant Physiol 151:1570–1581
33. Rai A, Saito K, Yamazaki M (2017) Integrated omics analysis of specialized metabolism in medicinal plants. Plant J 90:764–787
34. Dal'Molin CGO, Quek LE, Palfreyman RW, Brumbley SM, Nielsen LK (2010) AraGEM, a genome-scale reconstruction of the primary metabolic network in Arabidopsis. Plant Physiol 152:579–589
35. Pilalis E, Chatziioannou A, Thomasset B, Kolisis F (2011) An in silico compartmentalized metabolic model of *Brassica napus* enables the systemic study of regulatory aspects of plant central metabolism. Biotechnol Bioeng 108:1673–1682
36. Hay J, Schwender J (2011) Computational analysis of storage synthesis in developing *Brassica napus* L. (oilseed rape) embryos: flux variability analysis in relation to 13C metabolic flux analysis. Plant J 67:513–525
37. Hay J, Schwender J (2011) Metabolic network reconstruction and flux variability analysis of storage synthesis in developing oilseed rape (*Brassica napus* L.) embryos. Plant J 67:526–541
38. Grafahrend-Belau E, Schreiber F, Koschützki D, Junker BH (2009) Flux balance analysis of barley seeds: a computational approach to study systemic properties of central metabolism. Plant Physiol 149:585–598
39. Rolletschek H et al (2011) Combined noninvasive imaging and modeling approaches reveal metabolic compartmentation in the barley endosperm. Plant Cell 23:3041–3054
40. Dal'Molin CGO, Quek LE, Palfreyman RW, Brumbley SM, Nielsen LK (2010) C4GEM, a genome-scale metabolic model to study C4 plant metabolism. Plant Physiol 154:1871–1885
41. Saha R, Suthers PF, Maranas CD (2011) Zea mays irs1563: a comprehensive genome-scale metabolic reconstruction of maize metabolism. PLOS ONE 6:e21784
42. Poolman MG, Kundu S, Shaw R, Fell DA (2013) Responses to light intensity in a genome-scale model of rice metabolism. Plant Physiol 162:1060
43. Lakshmanan M et al (2015) Unraveling the light-specific metabolic and regulatory signatures of rice through combined in silico modeling and multiomics analysis. Plant Physiol 169:3002

44. Yuan H, Cheung CYM, Poolman Mark G, Hilbers Peter AJ, Riel Natal AW (2015) A genome-scale metabolic network reconstruction of tomato (*Solanum lycopersicum* L.) and its application to photorespiratory metabolism. Plant J 85:289–304
45. Soubeyrand E et al (2018) Constraint-based modeling highlights cell energy, redox status and α-ketoglutarate availability as metabolic drivers for anthocyanin accumulation in grape cells under nitrogen limitation. Front Plant Sci 9:421
46. Gomes de Oliveira Dal'Molin C, Quek L-E, Saa PA, Nielsen LK (2015) A multi-tissue genome-scale metabolic modeling framework for the analysis of whole plant systems. Front Plant Sci 6:4
47. Roslan ND et al (2012) Flavonoid biosynthesis genes putatively identified in the aromatic plant *Polygonum minus* via expressed sequences tag (EST) analysis. Int J Mol Sci 13:2692–2706
48. Loke K-K et al (2016) RNA-seq analysis for secondary metabolite pathway gene discovery in *Polygonum minus*. Genomics Data 7:12–13
49. Loke KK et al (2017) Transcriptome analysis of *Polygonum minus* reveals candidate genes involved in important secondary metabolic pathways of phenylpropanoids and flavonoids. Peer J 5:e2938
50. Ahmad R et al (2014) Volatile profiling of aromatic traditional medicinal plant, polygonum minus in different tissues and its biological activities. Molecules 19:19220–19242
51. Goh HH, Khairudin K, Sukiran NA, Normah MN, Baharum SN (2016) Metabolite profiling reveals temperature effects on the VOCs and flavonoids of different plant populations. Plant Biol 18:130–139
52. Hassim N et al (2015) Antioxidant and antibacterial assays on polygonum minus extracts: different extraction methods. Int J Chem Eng 2015:1–10
53. Ee SF et al (2013) Transcriptome profiling of genes induced by salicylic acid and methyl jasmonate in *Polygonum minus*. Mol Biol Rep 40:2231–2241
54. Rahnamaie-Tajadod R, Loke KK, Goh HH, Noor NM (2017) Differential gene expression analysis in *Polygonum minus* leaf upon 24h of methyl jasmonate elicitation. Front Plant Sci 8:109
55. Nazaruddin N et al (2017) Small RNA-seq analysis in response to methyl jasmonate and abscisic acid treatment in *Persicaria minor*. Genomics Data 12:157–158
56. Mehrotra B, Mendes P (2006) Biotechnology in agriculture and forestry, vol 57. Springer, Berlin/Heidelberg, pp 105–115
57. Aizat WM et al (2018) Extensive mass spectrometry proteomics data of *Persicaria minor* herb upon methyl jasmonate treatment. Data Brief 16:1091–1094

Chapter 6
Metabolic Engineering and Synthetic Biology

Ahmad Bazli Ramzi

Abstract In the modern era of next-generation genomics and Fourth Industrial Revolution, there is a growing demand for translational research that brings about not only impactful research but also potential commercialisation of R- and D-based products. Advancement of metabolic engineering and synthetic biology has put forward a viable and innovative biotechnological platform for bioproduct development especially using microbial chassis. In this chapter, readers will be introduced on the concepts of metabolic engineering, synthetic biology and microbial chassis and the applications of these biological engineering (BioE) components in the advancement of industrial and agricultural biotechnology. Main strategies in employing BioE platform are discussed especially for waste bioconversion and value-added product development. More importantly, this chapter will also discuss current endeavours in integrating systems and synthetic biology for microbial production of natural products by introducing flavonoid biosynthesis genes of *Polygonum minus*, a medicinally important tropical plant in engineered yeast.

Keywords Metabolic engineering · Synthetic biology · Microbial chassis · Biological engineering · Industrial biotechnology

6.1 Introduction

Genetic modification is central to global industrial practices be it in the traditional cross-breeding and mutagenesis means or recombinant DNA technology-inspired targeted genetic improvement. The common goal of these approaches is to improve biochemical reactions and obtain desirable traits or product titre from genetically modified organisms. Employment of genetic technologies is the hallmark of modern biotechnology that enables researchers to make changes at the DNA and protein levels for acquiring knowledge and creating new products and technologies to solve

A. B. Ramzi (✉)
Institute of Systems Biology, Universiti Kebangsaan Malaysia (UKM), Bangi, Malaysia
e-mail: bazliramzi@ukm.edu.my

W. M. Aizat et al. (eds.), *Omics Applications for Systems Biology*,
Advances in Experimental Medicine and Biology 1102,
https://doi.org/10.1007/978-3-319-98758-3_6

concurrent problems in healthcare, agriculture and environmental sectors. The advent of genomics and high-throughphut biology has accelerated the expansion of molecular biology tools and advanced genetic engineering techniques far-reaching than single-disciplinary recombinant protein expression and functional studies of individual gain-of-function and loss-of-function genetic mutants.

As discussed in the previous chapters, complex biological mechanisms elucidated using omics-driven systems biology platforms served as the basis and in-depth information on gene regulation, metabolic pathways and network modelling that enabled further applications and explorations using biotechnological- and engineering-based methodologies. Principally, metabolic engineering and synthetic biology are the expansion of the broad fields of molecular biology and represent the sophisticated versions of genetic engineering research areas that are built on recombinant DNA technology principles. The rapid progress of metabolic engineering, systems biology and synthetic biology is dovetailed with the emergence of revolutionary technologies such as RNA interference (RNAi) and clustered regularly interspaced short palindromic repeats (CRISPR)-based genome editing tools and booming bio-based industries and healthcare sectors in developed and developing countries. In this chapter, a particular emphasis is placed on the implementation of biological engineering (BioE) platform comprising of microbial-based metabolic engineering and synthetic biology approaches, as the enabling technology for integrating data-driven systems biology input essential in developing sustainable and value-added biotechnological applications.

6.1.1 Metabolic Engineering

The metabolic engineering era that started in the 1990s has led to many exciting research discoveries and accelerated an advanced genetic engineering approach that aimed towards investigating broader aspects of metabolic and biochemical interactions. Metabolic engineering is principally defined as the directed modulation of metabolic pathways using recombinant DNA technology for overproducing fuels, chemicals and pharmaceutical products, which generated greater attention and consequently rapid expansion due to its industrial relevance [1, 2]. Metabolic engineering somehow differs from the generic genetic engineering field with its virtue in the multilevel investigation of metabolic pathways and gene regulatory networks as compared to individual studies of genes and enzymes, especially the rate-limiting enzymes.

Fundamental underlying aspects of metabolic engineering involve genetic construction, performance analysis, pathway engineering and optimisation of metabolic pathways for attaining desirable products using techniques ranging from recombinant protein expression to biochemical analysis of flux, kinetics and thermodynamics of the engineered cells or proteins. In essentiality, metabolic engineering focuses on broader impacts of genetic modifications in the engineered cells by investigating stoichiometric balance, pathway regulation and network modelling, in systematic means that in a way precedes systems biology in understanding the biological mechanisms of complex metabolic perturbance at systems level [3]. This multidisciplinary research field has brought about seminal research findings such as the production of biofuels from amino acids [4], synthesis of antimalarial drug precursors [5], production of nonnatural chemicals such as 1,4-butanediol [6] and, more recently, complete biosynthesis of opiods using metabolically engineered microbes [7]. These landmark research have brought about emerging bio-based industries that employ metabolically engineered microbes for the production of biofuels, chemicals and pharmaceuticals with sustainable and green technology business models culminating in the creation of a new market segment for industrial biotechnology.

Metabolic engineers ultimately aim at improving production level while lowering energy burden and associated costs involved in product development and industrial commercialisation by debottlenecking, debugging and process optimisation of the engineered cells [8]. Greater understanding of the microbial genomes and metabolic pathways has accelerated genome-scale pathway engineering and systems metabolic engineering that utilise data-driven approaches including in silico and omics-based pathway prediction and gene selection tools for constructing, modulating and optimising of metabolic pathways and evolution of protein functions [9]. In silico-aided metabolic engineering approaches have sped up strain development for industrial production of amino acids and biochemicals that are attributed by the increased productivity and capacity of the engineered microbes for scale-up fermentation and bioprocessing [10]. The rapid progress of next-generation omics technologies and ever-expanding synthetic biology tools such as CRISPR interference (CRISPRi) shall further improve substrate utilisation and hyper-producing strain development via genome-wide analysis and high-throughput strain screening [11, 12].

6.1.2 Synthetic Biology

Synthetic biology is a rapidly emerging research discipline in the broad fields of molecular biology and has been interchangeably used in reference to metabolic engineering especially involving complicated genetic modification or alteration of living cells. The emergence of synthetic biology has been mostly associated with the lowering cost of high-throughput DNA synthesis and increasing interest in implementing engineering principles for modulating cellular and genetic systems. Synthetic biologists aim at standardising genetic tools and having greater defined control and modulation of the complex biological processes conferred by the genetic components based on abstraction hierarchy [13, 14]. Given the transdisciplinary aspects of this research field, the consensus definition of synthetic biology is the design and engineer of new biological parts, devices and systems as well as redesigning existing and natural biological system. In fact, synthetic biology shares overlapping characteristics of metabolic engineering particularly through the use of molecular biology and computational tools for DNA parts assembly, pathway engineering and use of well-studied model regulatory systems for genetic circuit designing and construction [2, 3]. Key defining events of this rapidly emerging field are dated back in the early 2000s where the first genetic counter and toggle switch were constructed and led to the development of various artificial genetic elements and control circuits including the use of logic gates for configurational control of gene expression [3, 15, 16]. Another important element of this field is the establishment of public collection and repositories such as Registry of Standard Biological Parts (RSBP), Synthetic Biology Open Language (SBOL) and Addgene public plasmid repositories that greatly aided parts standardisation and resource sharing among the scientific communities [17, 18].

Essential genetic elements such as promoter, ribosome binding site (RBS), coding sequence (CDS) and transcription terminator are considered as biological parts for constructing standardised biological systems with desired behaviour and purposes. The designing and construction of artificial biological systems with well-defined genetic components have led to the significant breakthrough in constituting synthetic bacterial genome using chemical DNA synthesis and catalytic DNA assembly techniques such as Gibson isothermal assembly and in vivo homologous recombination [19–21]. Big progress made in Synthetic Yeast 2.0, the international Synthetic Yeast Genome project that aims at redesigning and constructing a synthetic eukaryotic genome, has been greatly facilitated by de novo biodesign tools and smart data-intensive technologies which will mark another important achievement for advancing microbial synthetic biology [22, 23]. The employment of synthetic genetic systems will allow greater control of the biological functions and outputs that are important in generating genetic design automation and customisation based on iterative design-build-test-learn model cycle [24, 25]. Programming and bioinformatics tools are instrumental in implementing the iterative circuit cycle that allows systematic means of designing, simulating, predicting and analysing the overall research scheme that can be constantly improved in high-throughput manners as compared to the conventional genetic engineering methods of build and test individual constructs.

The synthetic biology approaches are markedly useful for the production of biochemicals with complex biosynthetic pathways and gene clusters such as plant secondary metabolites and antibiotics through computer-aided design (CAD) tools and biosensor-based approaches for designing, constructing and screening of important and rate-limiting enzymes of the targeted pathway. CAD tools such as antiSMASH 3.0 [26] and RetroPath [27] have been employed for predicting, screening and, ultimately, producing the targeted compound using microbial chassis [28]. Rapid strain development with enhanced production activity and robustness has been aided using genetically encoded biosensors that enable precise gene control and high-throughput screening of targeted transcriptional, enzymatic and cellular activities based on colorimetric and fluorescence signals of the genetic constructs in whole cell or cell-free formats [29, 30].

6.1.3 *Microbial Chassis for Metabolic Engineering and Synthetic Biology Endeavours*

A chassis in synthetic biology refers to the host organism that fundamentally supports the genetic system and provides the essential cellular resources such as transcription and translation machinery for the genetic component to function as intended. Advancement of metabolic engineering and synthetic biology has been built on the fundamental genetics and mathematical modelling of microbial chassis specifically, *Escherichia coli* and *Saccharomyces cerevisiae*, considered as two of the most important model microbes for prokaryotes and eukaryotes, respectively. Important features of these model microbes include highly amenable to genetic manipulation, well-developed genetic tools and in-depth genome information and genetic control system. Both systems have been widely used as standard biological systems with an ever-expanding repertoire of synthetic biology toolbox being developed for parts/device assembly, modulated gene control, pathway construction and biomanufacturing of industrially important products.

Apart from these microbes, Gram-positive *Bacillus subtilis* and *Corynebacterium glutamicum* have also been well-characterised and well-utilised for fundamental and synthetic biology applications with an increasing catalogue of characterised biological parts that are interchangeable using suitable DNA assembly methods such as BioBrick™ and isothermal assembly modules. In general, bacterial chassis offers a simple nutrient requirement, rapid growth and expression system, while yeast chassis provides a better secretory mechanism and post-translational modification benefits especially for the production of eukaryotic proteins. Well-established DNA delivery methods and a more standardised and improved genetic toolkit are increasingly available for these microbes using modular DNA assembly techniques such as ePathBrick [31] and CoryneBrick [32] in addition to the homologous recombination-based in vivo DNA assembly in yeast. Detailed characteristics and advantages of using these microbial chassis for biotechnological applications can be found in these excellent reviews [33, 34].

6.2 Applications of Metabolic Engineering and Synthetic Biology for Industrial and Agricultural Biotechnology

In this section, biotechnological applications of metabolic engineering and synthetic biology are presented and proposed as sustainable and 'greener' approaches in addressing perennial problems in industrial and agricultural sectors. BioE platform using model microbial chassis was devised and demonstrated for three main thrusts: biosynthesis of value-added products, bioconversion of industrial waste and improved system for bioremediation. This section provides the proof of concept on the applications and feasibility of employing microbial strains specifically *C. glutamicum*, *S. cerevisiae* and *E. coli* as BioE chassis that can be further developed to meet industrial and environmental demands.

6.2.1 Biosynthesis of Value-Added Products

Metabolic engineering and synthetic biology approaches have been instrumental in developing biotechnological products with increased productivity and promoting sustainability in the product development. There is an extensive repertoire of biochemicals, biofuels and pharmaceuticals that are biosynthesised in engineered microbes using natural and, more importantly, nonnatural substrates by introducing an array of genes and regulatory elements suited for the production of desired products. Plasmid construction and genetic designs are critical in pathway engineering for directing the overproduction of the targeted products without heavily affecting the engineered cell fitness. In our efforts to develop high-performing strains, there are several important considerations in devising biological engineering strategies which are outlined in Fig. 6.1.

Using the BioE platform, *C. glutamicum* has been successfully engineered to overproduce 5-aminolevulinic acid (ALA) by combining pathway engineering and metabolic perturbation workflow. ALA is a non-proteinogenic amino acid that is in high demand for its use as biofertiliser (agriculture), photosensitiser (medical) and acne treatment (cosmetics). For industrial applications, ALA is currently produced using non-sustainable chemical synthesis which implicates the overall costs of the product development. To this end, *C. glutamicum* bacterium, which naturally synthesises high amount of glutamate, was engineered to overproduce ALA using glutamate as the building block of the ALA backbone [35]. *C. glutamicum* is well-known as the industrial producer of amino acids with well-established pathways of TCA cycle at the pyruvate node and glutamate node.

To establish a high-performing expression system in *C. glutamicum*, a strong constitutive Trc promoter was used in place of the weak p-Out promoter and signal sequence of the original pMT1s expression plasmid [36]. The promoter sequence includes new RBS that was designed to be high in purines and within 7–9 bp distance from the CDS start codon. To convert glutamate to ALA, two genes from the C5 pathway, glutamyl-tRNA reductase HemA and glutamate 1-semialdehyde aminotransferase HemL were co-expressed in the engineered *C. glutamicum*. A feedback-deregulated HemA was created by introducing two lysine (KK) residues at the third position of HemA variants from several selected bacterial strains. Following ALA production screening, mutated HemA from *Salmonella typhimurium* was selected and co-expressed with *E. coli* BL21 (DE3) HemL on pMT-Trc plasmid yielding pHemAL construct. Addition of penicillin and 2,2′-dipyridyl was carried out to improve ALA production based on increasing glu-

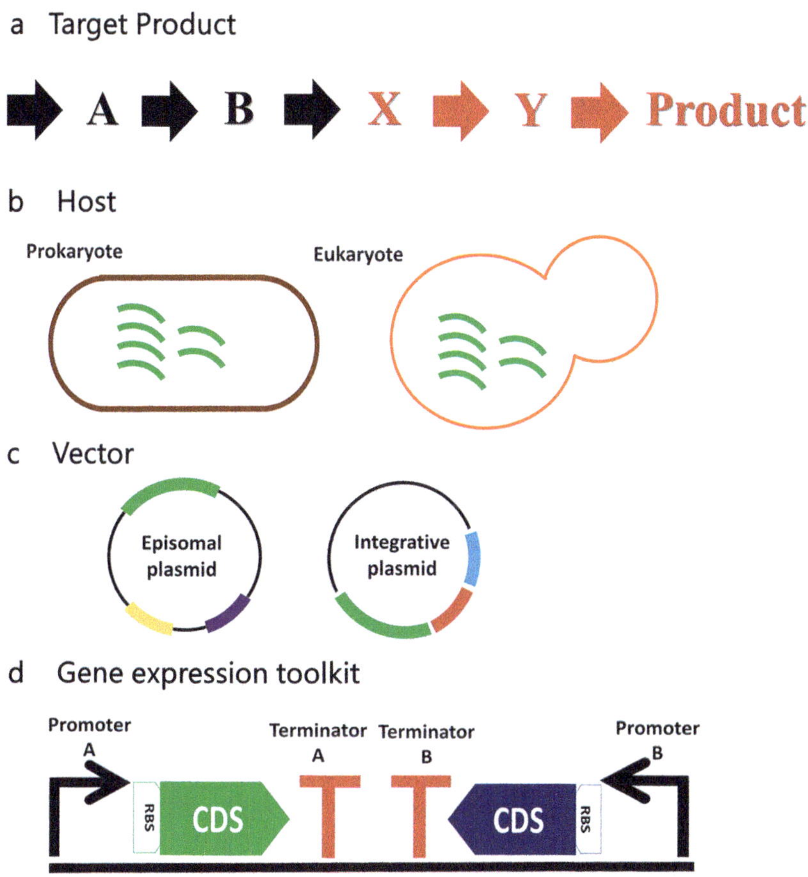

Fig. 6.1 Main aspects of biological engineering (BioE) strategies for value-added product development by engineered microbial hosts. (**a**) Selection of product and corresponding biosynthetic pathway/gene circuit. (**b**) Selection of prokaryotic and/or eukaryotic host. (**c**) Acquiring a suitable type of plasmid vector based on selection marker, replication mode and copy number. (**d**) Designing and constructing biological parts for desirable gene expression performance. Abbreviations: CDS coding sequence, RBS ribosome binding site

tamate flux and lowering HemA-limiting heme formation, respectively. In shake-flask fermentation, engineered *C. glutamicum* produced about 1.1 g/L ALA and 2.2 g/L in respective HemA and HemAL operon-expressing strains that represent 10.7-fold and 22-fold improvement from the control strain. Overall, the specially designed bioengineering strategies succeeded in improving ALA production in engineered *C. glutamicum* that can be used as the platform strain for higher-scale synthesis of ALA for potential commercial applications.

6.2.2 *Bioconversion of Agricultural Waste*

Metabolic engineering and synthetic biology are considered as the enabling technologies for waste-to-wealth concept especially through the development of bioengineered microbial strains. These powerful approaches have enabled and extended the range of carbon sources utilisation by engineered microbes using natural and nonnatural substrates mainly from cellulosic biomass waste made up of cellulose, hemicellulose and inhibitory amounts of acetic acid. Metabolic pathway designing often focuses on maintaining stoichiometric balance and cell growth of the microbial strains when grown on the waste substrates. Figure 6.2 illustrates an overview of the metabolic pathways and key enzymes involved in the utilisation of cellulosic biomass waste and glycerol waste using *S. cerevisiae* as the model microbe for waste-to-biofuel bioconversion.

Glycerol is a major by-product of the biodiesel industry and represents an inexpensive feedstock for bio-based product development. Bioconversion of waste by *S. cerevisiae* is of huge biotechnological interest due to the yeast bioprocessing capacity as an industrial workhorse for bioenergy (e.g. first-generation bioethanol) and biochemical (e.g. antimalarial drug precursor) production. However, wild-type yeast *S. cerevisiae* could not grow on glycerol as the sole substrate, hence preventing direct waste conversion using wild-type strains. Thus, to confer high-rate glycerol-utilising capability, *S. cerevisiae* YPH499 strain was engineered to overexpress glycerol dehydrogenase (Gcy) and dihydroxyacetone kinase (Dak), the key enzymes in converting glycerol to dihydroxyacetone phosphate, a glycolytic pathway intermediate in addition to the expression of glycerol uptake/transporter protein (Gup1) for improved glycerol uptake and utilisation [39, 40]. The engineered *S. cerevisiae* expressing Gcy, Dak and Gup1 formed the starting platform for glycerol bioconversion to bioethanol that was further enhanced via additional expression of alcohol dehydrogenase (Adh1), pyruvate decarboxylase (Pdc1), SAGA (Spt-Ada-Gcn5-acetyltransferase) complex genes and reduction of endogenous glycerol production Fps1 and Gpd2 genes [41, 42]. Overall, the specially designed bioengineering strategies successfully led to improved ethanol bioconversion up to 8.1 g/L of ethanol produced by engineered *S. cerevisiae* strains using glycerol as the main substrate.

Similar bioengineering strategies were employed for bioconversion of triacylglycerol (TAG), a microbial oil feedstock for biodiesel production using engineered *S. cerevisiae*. The overproduction of TAG from glycerol in *S. cerevisiae* was attained by introducing TAG biosynthetic genes diacylglycerol acyltransferase (DGA1) and phospholipid diacylglycerol acyltransferase (LRO1) in tandem with glycerol kinase (GUT1) for enhancing glycerol utilisation [43]. The engineered yeast produced about 8.2% overall TAG content, a 2.3-fold enhancement from the wild-type strain (3.6%). The microbial oil synthesised via bioengineering approaches offers an alternative and highly abundant nonfood feedstock for microdiesel and fatty acid-based third-generation biofuel development in the growing bioenergy sectors.

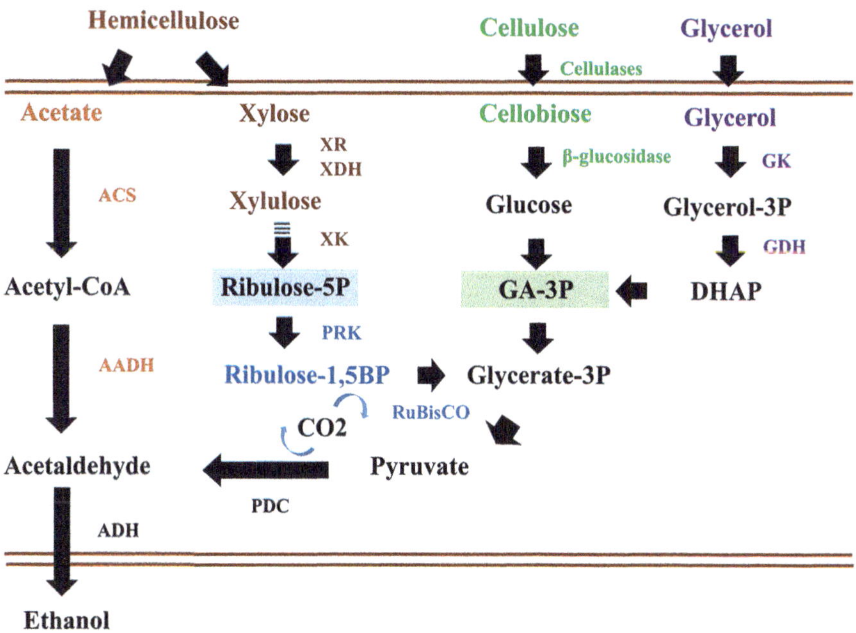

Fig. 6.2 Simplified scheme of metabolic pathways and key enzymes involved in waste-to-ethanol bioconversion by engineered *S. cerevisiae*. Nonnatural substrates and corresponding enzymes involved in the biocatalysis are highlighted in the same colour code. Acetate utilisation is coupled with xylose conversion from pretreated hemicellulosic waste, and cellulose bioconversion involves multiple enzymatic pretreatments and subsequent uptake of cellobiose extracellularly or intracellularly using transporter protein [37]. Glycerol utilisation requires corresponding glycerol uptake and conversion enzymes. An innovative approach that integrates carbon dioxide (CO_2) fixation with the endogenous ethanol fermentation pathway using non-native enzymes in xylose-fermenting *S. cerevisiae* is also highlighted [38]. Solid and dotted black arrows indicate metabolic reaction with single and multiple intermediates, respectively. Light blue and green boxes denote metabolic node for pentose phosphate (PP) and glycolytic pathway, respectively. Abbreviations for enzymes are as follows: ACS acetyl coenzyme A synthetase, AADH acetylating acetaldehyde dehydrogenase, XR xylose reductase, XDH xylitol dehydrogenase, XK xylulose kinase, PRK phosphoribulokinase, RuBisCO ribulose-1,5-bisphosphate carboxylase/oxygenase, GK (Gut1) glycerol kinase, GDH (Gcy) glycerol dehydrogenase, PDC pyruvate decarboxylase, and ADH, alcohol dehydrogenase. DHAP denotes dihydroxyacetone phosphate; P and BP abbreviate for intermediate compound with phosphate and biphosphate group, respectively

6.2.3 Improved System for Bioremediation

Modulation of metabolic pathways and genetic circuits is the cornerstone of metabolic engineering and synthetic biology approaches that provide researchers with the means for creating a finely tuned genetic system. Pathway engineering has been important in debottlenecking the targeted pathway for improving the flux and providing the precursors and cofactors required for optimal performance of the engineered cells. As depicted in Fig. 6.3, the overall performance of the genetic system

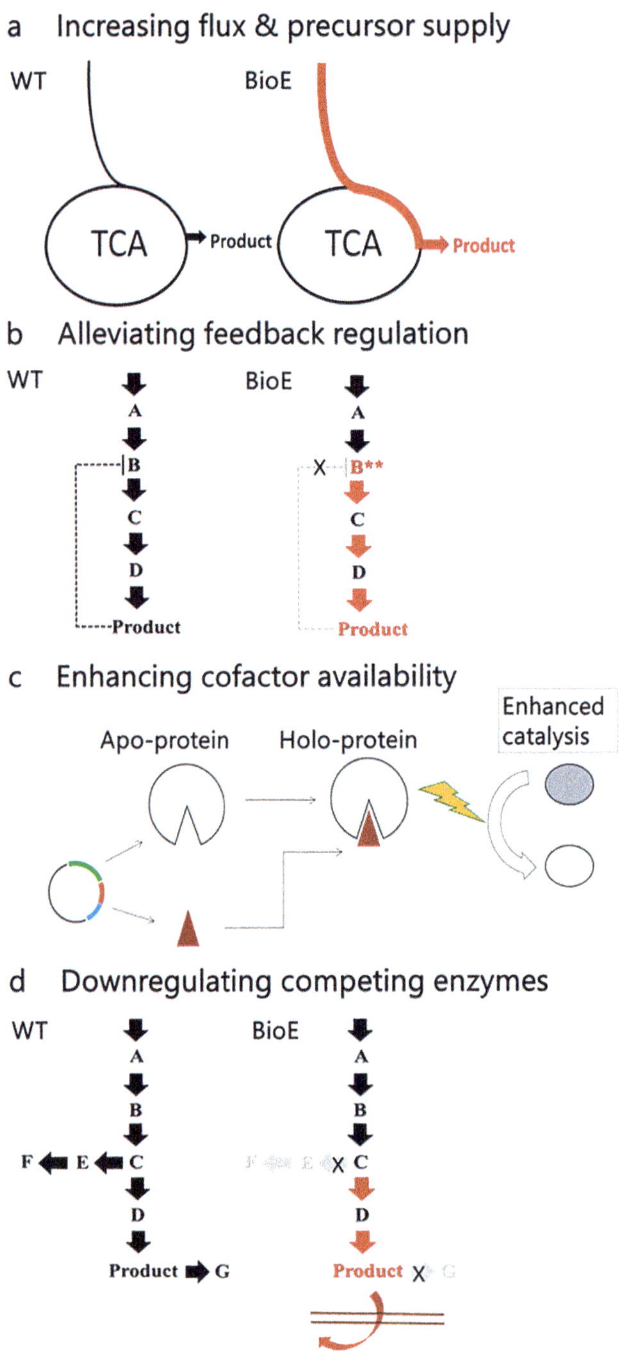

Fig. 6.3 Pathway engineering strategies for improving the performance of bioengineered cells and targeted output. (**a**) Increasing flux and precursor supply via gene expression enhancement. (**b**) Eliminating negative feedback regulation of rate-limiting enzyme via protein engineering or mutagenesis. (**c**) Increasing endogenous cofactor supply and regeneration via pathway modulation/optimisation. (**d**) Reducing the effects of competing enzymes via gene downregulation and extracellular product secretion. WT denotes for wild type and BioE represents bioengineered system

in bioengineered chassis can be improved using pathway engineering strategies that focused on increasing the catalytic activities of key enzymes in the targeted metabolic reactions.

Bioremediation is one of the promising applications of metabolic engineering and synthetic biology especially for degrading recalcitrant compounds in the environment. Natural microbes and enzymes have been extensively investigated for their uses in treating contaminants, but as often the case, process optimisation and associated costs hampered further in field applications. Biocatalysis by dye-decolourising peroxidase (DyP) is of great biotechnological interest due to its catalytic ability in degrading xenobiotics and lignin derivative compounds, hence offering a non-chemical approach for bioremediation and bioenergy applications. To date, development of DyP-based bioremediation is limited by the costly supply of precursors specifically ALA and heme chemicals that are important for functionality and catalytic activities of recombinant heme-containing DyP.

To this end, a bioengineering-inspired recombinant expression system was developed for producing recombinant DyP in engineered *E. coli* [44]. The focus was to enhance endogenous supply of heme cofactor for increasing the amount of holo-DyP protein via C5 biosynthesis pathway modulation. The need for exogenous precursor addition was offset by co-expressing a synthetic HemAL operon that increased ALA and heme content in the recombinant *E. coli*. The ALA-overproducing HemAL operon and recombinant Dyp from *Bacillus subtilis* were expressed using T7 expression systems to yield pHemAL-DyP plasmid construct. When compared with other systems, specifically DyP only and DyP added with hemin, the peroxidase activity of pHemAL-DyP was markedly increased of up to 66.7 U/mg in comparison with 39.0 and 43.4 U/mg peroxidase activity for pDyP and pDyP + hemin, respectively. Importantly, the pHemAL-DyP construct demonstrated an increased dye-decolourising activity by giving out the highest decolourisation percentage of 84.7% when tested on Reactive Blue 19 dye as compared to respective recombinant DyP only (69.9%) and DyP supplied with exogenous hemin (72.8%). The improved catalytic activities of the recombinant DyP will aid in developing crude DyP-based bioremediation of recalcitrant dyes and lignin residues in wastewater and other fields. Therefore, BioE-mediated pathway engineering represents a feasible and effective approach for improving genetic system performance and lowering the production costs by eliminating the need to supply exogenous chemical in the bacterial fermentation.

6.3 Case Study: Integrating Systems and Synthetic Biology for Omics-Driven Microbial Production of Natural Products

As discussed in the previous chapters, multidisciplinary systems biology research has been actively pursued in the context of multiple omics including transcriptomics, proteomics and metabolomics platforms. Using *P. minus* as the focal point, fundamental aspects of aromatic and bioactive properties of the medicinally important tropical herb were unravelled at molecular and systems levels [45, 46]. Despite its popular use in delicacy and folk medicine, potential commercialisation of single mixture and purified bioactive compounds of *P. minus* remains untapped due to the lack of enabling technologies other than conventional chemical extraction techniques that beset with poor yields and low productivity.

Consequently, metabolic engineering and synthetic biology approaches were undertaken to complement and utilise biological information gathered from the multi-omics analysis of *P. minus*. Considering the natural abundance and high antioxidant properties of the herbal extracts, flavonoid was chosen for its targeted production in bioengineered yeast by introducing a total of six biosynthesis genes for naringenin formation via native L-phenylalanine metabolic route. Figure 6.4 demonstrates a schematic representation of the microbial production of flavonoid using systems and synthetic biology research platforms.

Using *S. cerevisiae* as the microbial chassis, the integrative systems and synthetic research efforts were mainly aimed at synthesising bioactive natural products in the engineered yeast by introducing *P. minus* flavonoid biosynthetic pathway. Transcriptome dataset of *P. minus* was utilised in the mining of flavonoid biosynthetic genes including *PAL*, *C4H*, *CPR*, *4CL*, *CHS* and *CHI* to constitute a complete naringenin production pathway from endogenous L-phenylalanine in engineered *S. cerevisiae*. To facilitate a rapid construction of plasmid with multiple genes, a

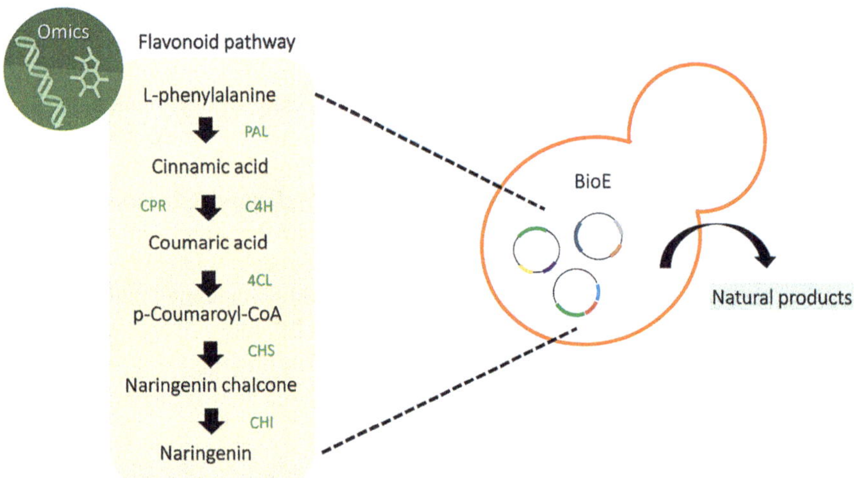

Fig. 6.4 Schematic representation of the microbial production of flavonoid using systems and synthetic biology approaches. Flavonoid biosynthetic pathway genes identified from transcriptomics analysis of *P. minus* were selected for heterologous pathway construction for natural products biosynthesis in engineered *S. cerevisiae*. Episomal plasmid vectors with different selection markers were employed in the pathway construction. BioE represents bioengineered yeast system. Abbreviations for enzymes are as follows: PAL phenylalanine ammonia lyase, C4H cinnamate-4-hydroxylase, CPR cytochrome P450 reductase, 4CL 4-coumarate-CoA ligase, CHS chalcone synthase, CHI chalcone isomerase

homology-based joining of rate-limiting enzyme C4H and yeast regulatory elements Tef1 promoter and ADH1 terminator was successfully carried out using isothermal assembly methods. Further plasmid construction was employed using modular Gateway and pCEV series plasmid vectors to direct the production of phenylpropanoid compounds in the engineered yeast strains. Successful implementation of this project shall represent an enabling technological platform for tropical plant-based natural product development via the integration of systems and synthetic biology approaches. In summary, further development of bioengineered systems for natural product biosynthesis using tropical genetic resources is envisaged owing to the rapid progression of data-driven high-throughput biologics and next-generation sequencing technologies that are available to systems and synthetic biology research communities.

References

1. Bailey JE (1991) Toward a science of metabolic engineering. Science 252:1668
2. Nielsen J, Keasling JD (2011) Synergies between synthetic biology and metabolic engineering. Nat Biotechnol 29:693
3. Stephanopoulos G (2012) Synthetic biology and metabolic engineering. ACS Synth Biol 1:514–525
4. Atsumi S, Hanai T, Liao JC (2008) Non-fermentative pathways for synthesis of branched-chain higher alcohols as biofuels. Nature 451:86
5. Ro DK et al (2006) Production of the antimalarial drug precursor artemisinic acid in engineered yeast. Nature 440:940–943
6. Yim H et al (2011) Metabolic engineering of *Escherichia coli* for direct production of 1,4-butanediol. Nat Chem Biol 7:445
7. Galanie S, Thodey K, Trenchard IJ, Interrante MF, Smolke CD (2015) Complete biosynthesis of opioids in yeast. Science 349:1095–1100
8. Keasling JD (2010) Manufacturing molecules through metabolic engineering. Science 330:1355
9. Chae TU, Choi SY, Kim JW, Ko Y-S, Lee SY (2017) Recent advances in systems metabolic engineering tools and strategies. Curr Opin Biotechnol 47:67–82
10. Lee SY, Kim HU (2015) Systems strategies for developing industrial microbial strains. Nat Biotechnol 33:1061
11. Mougiakos I, Bosma EF, Ganguly J, van der Oost J, van Kranenburg R (2018) Hijacking CRISPR-Cas for high-throughput bacterial metabolic engineering: advances and prospects. Curr Opin Biotechnol 50:146–157
12. Donohoue PD, Barrangou R, May AP (2018) Advances in industrial biotechnology using CRISPR-Cas systems. Trends Biotechnol 36:134–146
13. Endy D (2005) Foundations for engineering biology. Nature 438:449–453
14. Canton B, Labno A, Endy D (2008) Refinement and standardization of synthetic biological parts and devices. Nat Biotechnol 26:787–793
15. Elowitz MB, Leibler S (2000) A synthetic oscillatory network of transcriptional regulators. Nature 403:335–338
16. Gardner TS, Cantor CR, Collins JJ (2000) Construction of a genetic toggle switch in *Escherichia coli*. Nature 403:339–342
17. Heinemann M, Panke S (2006) Synthetic biology—putting engineering into biology. Bioinformatics 22:2790

18. Cameron DE, Bashor CJ, Collins JJ (2014) A brief history of synthetic biology. Nat Rev Microbiol 12:381–390
19. Gibson DG et al (2010) Creation of a bacterial cell controlled by a chemically synthesized genome. Science 329:52–56
20. Gibson DG et al (2008) One-step assembly in yeast of 25 overlapping DNA fragments to form a complete synthetic *Mycoplasma genitalium* genome. Proc Natl Acad Sci U S A 105:20404–20409
21. Gibson DG, Venter JG (2014) Synthetic biology: construction of a yeast chromosome. Nature 509:168–169
22. Richardson SM et al (2017) Design of a synthetic yeast genome. Science 355:1040
23. Pretorius IS, Boeke JD (2018) Yeast 2.0—Connecting the dots in the construction of the world's first functional synthetic eukaryotic genome. FEMS Yeast Res 18(4):foy032
24. Nielsen AA et al (2016) Genetic circuit design automation. Science 352:aac7341
25. Appleton E, Densmore D, Madsen C, Roehner N (2017) Needs and opportunities in bio-design automation: four areas for focus. Curr Opin Chem Biol 40:111
26. Weber T et al (2015) AntiSMASH 3.0—a comprehensive resource for the genome mining of biosynthetic gene clusters. Nucleic Acids Res 43:W237
27. Carbonell P, Parutto P, Baudier C, Junot C, Faulon JL (2014) Retropath: automated pipeline for embedded metabolic circuits. ACS Synth Biol 3:565
28. Fehér T et al (2014) Validation of RetroPath, a computer-aided design tool for metabolic pathway engineering. Biotechnol J 9:1446
29. Rogers JK, Taylor ND, Church GM (2016) Biosensor-based engineering of biosynthetic pathways. Curr Opin Biotechnol 42:84–91
30. Jiang L, Zhao J, Lian J, Xu Z (2018) Cell-free protein synthesis enabled rapid prototyping for metabolic engineering and synthetic biology. Synth Sys Biotechnol 3:90–96
31. Xu P, Vansiri A, Bhan N, Koffas MAG (2012) EPathBrick: a synthetic biology platform for engineering metabolic pathways in *E. coli*. ACS Synth Biol 1:256
32. Kang MK et al (2014) Synthetic biology platform of CoryneBrick vectors for gene expression in *Corynebacterium glutamicum* and its application to xylose utilization. Appl Microbiol Biotechnol 98:5991
33. Kavšček M, Stražar M, Curk T, Natter K, Petrovič U (2015) Yeast as a cell factory: current state and perspectives. Microb Cell Factories 14:94
34. Calero P, Nikel PI (2018) Chasing bacterial *chassis* for metabolic engineering: a perspective review from classical to non-traditional microorganisms. Microb Biotechnol [Epub ahead of print]
35. Ramzi AB, Hyeon JE, Kim SW, Park C, Han SO (2015) 5-Aminolevulinic acid production in engineered *Corynebacterium glutamicum* via C5 biosynthesis pathway. Enzyme Microb Technol 81:1–7
36. Hyeon JE, Jeon WJ, Whang SY, Han SO (2011) Production of minicellulosomes for the enhanced hydrolysis of cellulosic substrates by recombinant *Corynebacterium glutamicum*. Enzyme Microb Technol 48:371
37. Wei N, Oh EJ, Million G, Cate JHD, Jin YS (2015) Simultaneous utilization of cellobiose, xylose, and acetic acid from lignocellulosic biomass for biofuel production by an engineered yeast platform. ACS Synth Biol 4:707
38. Xia PF et al (2016) Recycling carbon dioxide during xylose fermentation by engineered *Saccharomyces cerevisiae*. ACS Synth Biol 6:276–283
39. Yu KO, Kim SW, Han SO (2010) Engineering of glycerol utilization pathway for ethanol production by *Saccharomyces cerevisiae*. Bioresour Technol 101:4157–4161
40. Yu KO, Kim SW, Han SO (2010) Reduction of glycerol production to improve ethanol yield in an engineered *Saccharomyces cerevisiae* using glycerol as a substrate. J Biotechnol 150:209–214
41. Yu KO et al (2012) Increased ethanol production from glycerol by *Saccharomyces cerevisiae* strains with enhanced stress tolerance from the overexpression of SAGA complex components. Enzyme Microb Technol 51:237–243

42. Yu KO et al (2012) Improvement of ethanol yield from glycerol via conversion of pyruvate to ethanol in metabolically engineered *Saccharomyces cerevisiae*. Appl Biochem Biotechnol 166:856–865
43. Yu KO et al (2013) Development of a *Saccharomyces cerevisiae* strain for increasing the accumulation of triacylglycerol as a microbial oil feedstock for biodiesel production using glycerol as a substrate. Biotechnol Bioeng 110:343–347
44. Ramzi AB, Hyeon JE, Han SO (2015) Improved catalytic activities of a dye-decolorizing peroxidase (DyP) by overexpression of ALA and heme biosynthesis genes in *Escherichia coli*. Process Biochem 50:1272–1276
45. Ahmad R et al (2018) Polygonumins A, a newly isolated compound from the stem of *Polygonum minus* Huds with potential medicinal activities. Sci Rep 8(4202):4202
46. Loke K-K et al (2017) Transcriptome analysis of *Polygonum minus* reveals candidate genes involved in important secondary metabolic pathways of phenylpropanoids and flavonoids. Peer J 5:e2938

Index

W. M. Aizat et al. (eds.), *Omics Applications for Systems Biology*, Advances in Experimental Medicine and Biology 1102,
https://doi.org/10.1007/978-3-319-98758-3

MIX
Papier aus verantwortungsvollen Quellen
Paper from responsible sources
FSC® C105338

Printed by Books on Demand, Germany

MIX
Papier aus verantwortungsvollen Quellen
Paper from responsible sources
FSC® C105338

Printed by Books on Demand, Germany